98-CTS-2

USE OF NOVEL TECHNIQUES TO QUANTIFY PHENOTYPES IN BIOLOGICAL TREATMENT PROCESSES

AF621713

by:

Kevin G. Robinson
Gary S. Sayler
Janalyn R. Brown
Jack M. Carley
Matthew Derrigo
Hebe M. Dionisi
Jason Flowers
Victoria M. Garrett
Igrid Gregory
Gerda Harms
Shawn A. Hawkins
Alice C. Layton
Jack J. Parker
Brent C. Wood
University of Tennessee

Arthur J. Meyers
Eastman Chemical Company

2004

WERF

The Water Environment Research Foundation, a not-for-profit organization, funds and manages water quality research for its subscribers through a diverse public-private partnership between municipal utilities, corporations, academia, industry, and the federal government. WERF subscribers include municipal and regional water and wastewater utilities, industrial corporations, environmental engineering firms, and others that share a commitment to cost-effective water quality solutions. WERF is dedicated to advancing science and technology addressing water quality issues as they impact water resources, the atmosphere, the lands, and quality of life.

For more information, contact:
Water Environment Research Foundation
635 Slaters Lane, Suite 300
Alexandria, VA 22314-1177
Tel: (703) 684-2470
Fax: (703) 299-0742
www.werf.org
werf@werf.org

This report was co-published by the following organizations. For non-subscriber sales information, contact:

IWA Publishing
Alliance House, 12 Caxton Street
London SW1H 0QS, United Kingdom
Tel: +44 (0) 20 7654 5500
Fax: +44 (0) 20 7654 5555
www.iwapublishing.com
publications@iwap.co.uk

© Copyright 2004 by the Water Environment Research Foundation. All rights reserved. Permission to copy must be obtained from the Water Environment Research Foundation.
Library of Congress Catalog Card Number: 2003109084

IWAP ISBN: 1-84339-697-1

This report was prepared by the organization(s) named below as an account of work sponsored by the Water Environment Research Foundation (WERF). Neither WERF, members of WERF, the organization(s) named below, nor any person acting on their behalf: (a) makes any warranty, express or implied, with respect to the use of any information, apparatus, method, or process disclosed in this report or that such use may not infringe on privately owned rights; or (b) assumes any liabilities with respect to the use of, or for damages resulting from the use of, any information, apparatus, method, or process disclosed in this report.

University of Tennessee, Eastman Chemical Company

This document was reviewed by a panel of independent experts selected by WERF. Mention of trade names or commercial products does not constitute WERF nor EPA endorsement or recommendations for use. Similarly, omission of products or trade names indicates nothing concerning WERF's nor EPA's positions regarding product effectiveness or applicability.

The research on which this report is based was funded, in part, by the United States Environmental Protection Agency through Cooperative Agreement No. CR-825237-01 with the Water Environment Research Foundation (WERF). Unless an EPA logo appears on the cover, this report is a publication of WERF, not EPA. Funds awarded under the Cooperative Agreement cited above were not used for editorial services, reproduction, printing, or distribution.

ACKNOWLEDGMENTS

The project team would like to acknowledge the University of Tennessee Waste Management Research and Education Institute for partial funding of this work. Additionally, we thank Eastman Chemical Company for use of laboratory facilities and the Knoxville Utilities Board for providing municipal samples. We also thank Neil Quigley at the Molecular Biology Resource Facility (University of Tennessee, Knoxville) for DNA sequencing.

Report Preparation

Principal Investigator:
Kevin G. Robinson, Ph.D., MSPH
University of Tennessee

Project Team:
Gary S. Sayler, Ph.D.
Janalyn R. Brown, M.S.
Jack M. Carley, M.S.
Matthew Derrigo, B.S.
Hebe M. Dionisi, Ph.D.
Jason Flowers, B.S.
Victoria M. Garrett, M.S.
Igrid Gregory, M.S.
Gerda Harms, Ph.D.
Shawn A. Hawkins, M.S., P.E.
Alice C. Layton, Ph.D.
Jack J. Parker, M.S.
Brent C. Wood, M.S.
University of Tennessee

Arthur J. Meyers, Ph.D.
Eastman Chemical Company

Project Subcommittee

Michael Aitken, Ph.D.
University of North Carolina

Steven Chiesa, Ph.D., P.E.
Santa Clara University

Glen Daigger, Ph.D., P.E., DEE
CH2M HILL

Charles F. Kulpa Jr., Ph.D.
University of Notre Dame

Vasantha Nagarajan
Dupont Engineering Company

Mesut Sezgin, Ph.D., P.E.
City of Atlanta, Department of Public Works

Water Environment Research Foundation Staff

Director of Research: Daniel M. Woltering, Ph.D.
Project Manager: Lola Olabode, M.P.H.

ABSTRACT AND BENEFITS

Abstract:

This research attempts to evaluate nitrification treatment performance in combined carbon/nitrogen municipal wastewater reactors using traditional physical/chemical methods and modern molecular techniques. Bench scale activated sludge reactors were operated at different solids retention times (SRT) with varying dissolved oxygen (DO) levels and temperatures over a 21-month period. Real-time PCR assays were used to determine cell concentrations of total bacterial 16S ribosomal deoxyribonucleic acid (rDNA), a gross measure of biomass content, the *amo*A gene, a measure of ammonia-oxidizing bacteria (AOB), and the *Nitrospira* 16S rDNA gene, a measure of nitrite-oxidizing bacteria (NOB). As expected, gravimetric biomass and total bacterial 16S rDNA levels increased with increasing SRT. Ammonia oxidation rates and *N. oligotropha*-type AOB concentrations did not follow similar trends with respect to changes in SRT, temperature, and DO; nor were they highly correlated. The concentration of available nitrite and SRT were positively correlated with *Nitrospira* cell densities; while DO concentration and temperature were negatively correlated with NOB levels. The percentage of the total population comprised of AOB and NOB obtained with the real-time polymerase chain reaction (PCR) assays was compared to predicted values estimated from design equations using typical kinetic parameters. While the percentages of NOB measured using the real-time PCR assay corresponded very well with the predicted values, the measured percentages of AOB were much lower than those estimated from the design equations, suggesting that *N. oligotropha*-type AOB were not the dominant ammonia-oxidizing species in these reactors.

Benefits:

- Develops a molecular method for use in "dirty" wastewater environments.
- Demonstrates that a nonconventional technique can be used to quantify specific groups of bacteria during nitrification treatment.
- Presents a new real-time PCR procedure that can be used to assess nitrifying populations during biological treatment.

Keywords: nitrification, real-time PCR, *amo*A, *Nitrospira*, 16S rDNA

TABLE OF CONTENTS

LIST OF TABLES

LIST OF FIGURES

LIST OF ACRONYMS

A	adenine
AMO	ammonia monooxygenase enzyme
amo	ammonia monooxygenase gene
amoA	ammonia monooxygenase gene subunit A
ANOVA	analysis of variance
AOB	ammonia-oxidizing bacteria
AODC	acridine orange direct counts
AOE	ammonia oxidation efficiency
APHA	American Public Health Association
BNR	biological nutrient removal
BOD	biochemical oxygen demand
BOD_5	five-day BOD
C	cytosine
COD	chemical oxygen demand
COD_s	soluble COD
COD_t	total COD
cPCR	competitive PCR
CSTR	continuous-stirred tank reactor
DDE	dynamic data exchange
DGGE	denaturing-gradient gel electrophoresis
DI	deionized
DNA	deoxyribonucleic acid
dNTP	deoxynucleotide triphosphates
DO	dissolved oxygen
EPA	Environmental Protection Agency
ESS	effluent suspended solids
FAM	6-carboxy-fluorescein
FISH	fluorescence in situ hybridization
G	guanine
HAO	hydroxylamine oxidoreductase enzyme
hao	hydroxylamine oxidoreductase gene
HART	Highway Addressable Remote Transducer
HPLC	high pressure liquid chromatography
HRT	hydraulic retention time
IAWPRC	International Association on Water Pollution Research and Control
IC	ion chromatograph
KHP	potassium hydrogen phthalate
KUB	Knoxville Utilities Board
MBR	membrane bioreactor
MLSS	mixed liquor suspended solids
MLVSS	mixed liquor volatile suspended solids
MPN	most probable number
mRNA	messenger RNA
N	nitrogen
NOB	nitrite-oxidizing bacteria

NOE	nitrogen oxidation efficiency
NOR	nitrite oxidoreductase enzyme
PC	personal computer
PCR	polymerase chain reaction
rDNA	ribosomal DNA
RNA	ribonucleic acid
rRNA	ribosomal RNA
SBR	sequencing batch reactor
SD	standard deviation
SRT	solids retention time
SVI	sludge volume index
T	thymine
TAMRA	6-carboxy-tetramethyl rhodamine
TBE	tris borate
TDPH	Tennessee Department of Public Health
TKN	total Kjeldahl nitrogen
UDG	uracil DNA glycosylase
VSS	volatile suspended solids
WWTP	wastewater treatment plant

98-CTS-2

USE OF NOVEL TECHNIQUES TO QUANTIFY PHENOTYPES IN BIOLOGICAL TREATMENT PROCESSES

by:

Kevin G. Robinson
Gary S. Sayler
Janalyn R. Brown
Jack M. Carley
Matthew Derrigo
Hebe M. Dionisi
Jason Flowers
Victoria M. Garrett
Igrid Gregory
Gerda Harms
Shawn A. Hawkins
Alice C. Layton
Jack J. Parker
Brent C. Wood
University of Tennessee

Arthur J. Meyers
Eastman Chemical Company

2004

WERF

The Water Environment Research Foundation, a not-for-profit organization, funds and manages water quality research for its subscribers through a diverse public-private partnership between municipal utilities, corporations, academia, industry, and the federal government. WERF subscribers include municipal and regional water and wastewater utilities, industrial corporations, environmental engineering firms, and others that share a commitment to cost-effective water quality solutions. WERF is dedicated to advancing science and technology addressing water quality issues as they impact water resources, the atmosphere, the lands, and quality of life.

For more information, contact:
Water Environment Research Foundation
635 Slaters Lane, Suite 300
Alexandria, VA 22314-1177
Tel: (703) 684-2470
Fax: (703) 299-0742
www.werf.org
werf@werf.org

This report was co-published by the following organizations. For non-subscriber sales information, contact:

IWA Publishing
Alliance House, 12 Caxton Street
London SW1H 0QS, United Kingdom
Tel: +44 (0) 20 7654 5500
Fax: +44 (0) 20 7654 5555
www.iwapublishing.com
publications@iwap.co.uk

© Copyright 2004 by the Water Environment Research Foundation. All rights reserved. Permission to copy must be obtained from the Water Environment Research Foundation.
Library of Congress Catalog Card Number: 2003109084
Printed in the United States of America
IWAP ISBN: 1-84339-697-1

This report was prepared by the organization(s) named below as an account of work sponsored by the Water Environment Research Foundation (WERF). Neither WERF, members of WERF, the organization(s) named below, nor any person acting on their behalf: (a) makes any warranty, express or implied, with respect to the use of any information, apparatus, method, or process disclosed in this report or that such use may not infringe on privately owned rights; or (b) assumes any liabilities with respect to the use of, or for damages resulting from the use of, any information, apparatus, method, or process disclosed in this report.

University of Tennessee, Eastman Chemical Company

This document was reviewed by a panel of independent experts selected by WERF. Mention of trade names or commercial products does not constitute WERF nor EPA endorsement or recommendations for use. Similarly, omission of products or trade names indicates nothing concerning WERF's nor EPA's positions regarding product effectiveness or applicability.

The research on which this report is based was funded, in part, by the United States Environmental Protection Agency through Cooperative Agreement No. CR-825237-01 with the Water Environment Research Foundation (WERF). Unless an EPA logo appears on the cover, this report is a publication of WERF, not EPA. Funds awarded under the Cooperative Agreement cited above were not used for editorial services, reproduction, printing, or distribution.

ACKNOWLEDGMENTS

The project team would like to acknowledge the University of Tennessee Waste Management Research and Education Institute for partial funding of this work. Additionally, we thank Eastman Chemical Company for use of laboratory facilities and the Knoxville Utilities Board for providing municipal samples. We also thank Neil Quigley at the Molecular Biology Resource Facility (University of Tennessee, Knoxville) for DNA sequencing.

Report Preparation

Principal Investigator:
Kevin G. Robinson, Ph.D., MSPH
University of Tennessee

Project Team:
Gary S. Sayler, Ph.D.
Janalyn R. Brown, M.S.
Jack M. Carley, M.S.
Matthew Derrigo, B.S.
Hebe M. Dionisi, Ph.D.
Jason Flowers, B.S.
Victoria M. Garrett, M.S.
Igrid Gregory, M.S.
Gerda Harms, Ph.D.
Shawn A. Hawkins, M.S., P.E.
Alice C. Layton, Ph.D.
Jack J. Parker, M.S.
Brent C. Wood, M.S.
University of Tennessee

Arthur J. Meyers, Ph.D.
Eastman Chemical Company

Project Subcommittee

Michael Aitken, Ph.D.
University of North Carolina

Steven Chiesa, Ph.D., P.E.
Santa Clara University

Glen Daigger, Ph.D., P.E., DEE
CH2M HILL

Charles F. Kulpa Jr., Ph.D.
University of Notre Dame

Vasantha Nagarajan
Dupont Engineering Company

Mesut Sezgin, Ph.D., P.E.
City of Atlanta, Department of Public Works

Water Environment Research Foundation Staff

Director of Research: Daniel M. Woltering, Ph.D.
Project Manager: Lola Olabode, M.P.H.

ABSTRACT AND BENEFITS

Abstract:

This research attempts to evaluate nitrification treatment performance in combined carbon/nitrogen municipal wastewater reactors using traditional physical/chemical methods and modern molecular techniques. Bench scale activated sludge reactors were operated at different solids retention times (SRT) with varying dissolved oxygen (DO) levels and temperatures over a 21-month period. Real-time PCR assays were used to determine cell concentrations of total bacterial 16S ribosomal deoxyribonucleic acid (rDNA), a gross measure of biomass content, the *amo*A gene, a measure of ammonia-oxidizing bacteria (AOB), and the *Nitrospira* 16S rDNA gene, a measure of nitrite-oxidizing bacteria (NOB). As expected, gravimetric biomass and total bacterial 16S rDNA levels increased with increasing SRT. Ammonia oxidation rates and *N. oligotropha*-type AOB concentrations did not follow similar trends with respect to changes in SRT, temperature, and DO; nor were they highly correlated. The concentration of available nitrite and SRT were positively correlated with *Nitrospira* cell densities; while DO concentration and temperature were negatively correlated with NOB levels. The percentage of the total population comprised of AOB and NOB obtained with the real-time polymerase chain reaction (PCR) assays was compared to predicted values estimated from design equations using typical kinetic parameters. While the percentages of NOB measured using the real-time PCR assay corresponded very well with the predicted values, the measured percentages of AOB were much lower than those estimated from the design equations, suggesting that *N. oligotropha*-type AOB were not the dominant ammonia-oxidizing species in these reactors.

Benefits:

- Develops a molecular method for use in "dirty" wastewater environments.
- Demonstrates that a nonconventional technique can be used to quantify specific groups of bacteria during nitrification treatment.
- Presents a new real-time PCR procedure that can be used to assess nitrifying populations during biological treatment.

Keywords: nitrification, real-time PCR, *amo*A, *Nitrospira*, 16S rDNA

TABLE OF CONTENTS

LIST OF TABLES

LIST OF FIGURES

LIST OF ACRONYMS

A	adenine
AMO	ammonia monooxygenase enzyme
amo	ammonia monooxygenase gene
amoA	ammonia monooxygenase gene subunit A
ANOVA	analysis of variance
AOB	ammonia-oxidizing bacteria
AODC	acridine orange direct counts
AOE	ammonia oxidation efficiency
APHA	American Public Health Association
BNR	biological nutrient removal
BOD	biochemical oxygen demand
BOD_5	five-day BOD
C	cytosine
COD	chemical oxygen demand
COD_s	soluble COD
COD_t	total COD
cPCR	competitive PCR
CSTR	continuous-stirred tank reactor
DDE	dynamic data exchange
DGGE	denaturing-gradient gel electrophoresis
DI	deionized
DNA	deoxyribonucleic acid
dNTP	deoxynucleotide triphosphates
DO	dissolved oxygen
EPA	Environmental Protection Agency
ESS	effluent suspended solids
FAM	6-carboxy-fluorescein
FISH	fluorescence in situ hybridization
G	guanine
HAO	hydroxylamine oxidoreductase enzyme
hao	hydroxylamine oxidoreductase gene
HART	Highway Addressable Remote Transducer
HPLC	high pressure liquid chromatography
HRT	hydraulic retention time
IAWPRC	International Association on Water Pollution Research and Control
IC	ion chromatograph
KHP	potassium hydrogen phthalate
KUB	Knoxville Utilities Board
MBR	membrane bioreactor
MLSS	mixed liquor suspended solids
MLVSS	mixed liquor volatile suspended solids
MPN	most probable number
mRNA	messenger RNA
N	nitrogen
NOB	nitrite-oxidizing bacteria

NOE	nitrogen oxidation efficiency
NOR	nitrite oxidoreductase enzyme
PC	personal computer
PCR	polymerase chain reaction
rDNA	ribosomal DNA
RNA	ribonucleic acid
rRNA	ribosomal RNA
SBR	sequencing batch reactor
SD	standard deviation
SRT	solids retention time
SVI	sludge volume index
T	thymine
TAMRA	6-carboxy-tetramethyl rhodamine
TBE	tris borate
TDPH	Tennessee Department of Public Health
TKN	total Kjeldahl nitrogen
UDG	uracil DNA glycosylase
VSS	volatile suspended solids
WWTP	wastewater treatment plant

Executive Summary

ES-1 Project Objectives

A thorough understanding of microbial dynamics in activated sludge treatment is currently under-realized. This is due, in part, to quantitative techniques that rely on culture-dependent methods. The bias inherent with cultivation techniques may result in isolates that do not necessarily represent those dominant in the system evaluated. In-situ analysis of microorganisms functioning in wastewater reactors will provide a greater understanding of how particular groups of organisms, such as nitrifiers, are impacted by system fluctuations. This information is a key to improving process control and process design as well as for the development of more efficient biotreatment strategies.

The overall goal of this research was to elucidate the number of specific nitrogen (N)-utilizing bacteria under varying operational and environmental conditions during activated sludge treatment using molecular-based techniques. A primary objective of this research was to evaluate a new molecular technique (real-time polymerase chain reaction [PCR]) as a tool to quantify nitrifying bacteria in activated sludge by developing PCR assays for 1) the eubacterial 16S ribosomal deoxyribonucleic acid (rDNA) gene (a measure of total bacteria present), 2) the *Nitrosomonas oligotropha-* and *Nitrosomonas nitrosa*-type *amoA* gene (a measure of ammonia-oxidizing bacteria present), and 3) the 16S rDNA gene from *Nitrospira* (a measure of nitrite-oxidizing bacteria present). Additionally, this research attempted to utilize these new PCR assays, in addition to traditional physical/chemical methods, to evaluate nitrification treatment performance in combined carbon/nitrogen activated sludge systems that treat municipal or industrial wastewater.

ES-2 Organization of Report

This report includes a review of recent literature relative to nitrification in single-stage activated sludge wastewater treatment systems (Chapter 2.0); molecular methods to assess treatment performance (Chapter 2.0); a description of procedures and methods utilized in this study (Chapter 3.0); a description of results from bench-scale operation of the municipal and industrial treatment systems and from molecular analysis of mixed liquor samples collected from those systems under various operating conditions (Chapter 4.0); and a summary of results and conclusions (Chapter 5.0).

ES-3 Research Approach

Bench-scale systems (with solids recycle) were developed and operated to treat municipal or industrial wastewater collected from two full-scale wastewater treatment plants (WWTPs). Experimental systems consisted of multiple continuous stirred tank reactors (CSTRs) (10 L each) operating in parallel. Each CSTR had an external secondary clarifier, designed to simulate the full-scale activated sludge treatment facilities. Impacts of operational and environmental parameters (solids retention time [SRT], dissolved oxygen [DO] and temperature) on nitrification performance and microbial dynamics were assessed. Each bench-scale reactor system was operated at 20-, 10-, 5-, and 2-day SRTs. For the municipal wastewater treatment system, four activated sludge reactors were operated at three constant temperature levels (20°C, 15°C, and 10°C) and a minimum of three DO levels: [high (≥2.0 mg/L), intermediate (1.0 mg/L to 1.5 mg/L) and low (≤0.5 mg/L)]. A control system was used to maintain DO target set points. For the industrial wastewater treatment system, four separate tanks, each consisting of three 10-L baffled reactors operating in series, were utilized. Each of the four treatment trains operated at a unique SRT. This system operated at two DO levels (high and low) and two ammonia-loading levels (high and low). All testing was performed at room temperature. In addition to a DO control system, a pH control system was utilized in the industrial treatment system.

Nitrification performance was determined by traditional physical/chemical measures and compared with cell numbers of key populations determined by real-time PCR under the different operating regimes. Total bacteria, ammonia-oxidizing bacteria, and nitrite-oxidizing bacteria were monitored using real-time PCR assays designed and optimized for four targets: bacterial 16S rDNA, *Nitrospira* 16S rDNA, *N. oligotropha*-type *amo*A, and *N. nitrosa*-type *amo*A.

Municipal Wastewater Treatment

Total Bacterial 16S rDNA

Tables ES-1, ES-2, and ES-3 show the mixed liquor volatile suspended solids (MLVSS) concentrations, the total bacterial 16S rDNA cell concentrations (normalized to mixed liquor volume), and the chemical oxygen demand (COD) removal efficiencies for the 20°C, 15°C, and 10°C systems, respectively. Results indicate that total bacterial concentrations follow the expected trends, with the highest average cell concentrations present in the 20-day SRT samples and the lowest average cell concentrations present in the 2-day SRT samples. While total bacterial concentration was found to be highly dependent on SRT, significant correlations with temperature or DO levels were not generally observed. Temperature and DO level would not be expected to have a significant impact on the total bacterial concentration in the range of operating conditions tested. Both of these parameters are largely reflective of the heterotrophic population, which is not as sensitive to changes in these conditions as the nitrifying bacteria. A comparison of average total bacterial 16S rDNA and MLVSS concentrations, as a function of SRT, shows that both measures of biomass followed similar trends with respect to this operational parameter.

Nitrosomonas oligotropha-type Ammonia Oxidizing Bacteria

Average influent and effluent ammonia concentration, and real-time PCR data for the *N. oligotropha*-type *amo*A gene (a measure of ammonia oxidizing bacteria [AOB] present) are presented in Tables ES-4, ES-5 and ES-6 for the 20°C, 15°C, and 10°C operating periods, respectively. AOB values were calculated assuming that *N. oligotropha*-type bacteria have two copies of the *amo*A gene per cell. The *N. oligotropha*-type AOB concentration tended to decrease with decreasing SRT during

Table ES-1. Average MLVSS (mg/L), Total Bacterial 16S rDNA Concentrations (cells/L), and COD Removal Efficiencies (%) During 20°C Operation.

SRT	High DO			Intermediate DO			Low DO		
	MLVSS	16S rDNA	COD Eff.	MLVSS	16S rDNA	COD Eff.	MLVSS	16S rDNA	COD Eff.
20-Day	2113	11.0×10^{11}	87%	2941	6.2×10^{11}	91%	2515	7.1×10^{11}	88%
10-Day	1238	7.0×10^{11}	86%	2041	5.2×10^{11}	89%	1686	6.5×10^{11}	87%
5-Day	644	3.8×10^{11}	83%	932	2.0×10^{11}	87%	854	1.8×10^{11}	85%
2-Day	358	1.3×10^{11}	82%	407	0.75×10^{11}	83%	345	1.1×10^{11}	81%

Table ES-2. Average MLVSS (mg/L), Total Bacterial 16S rDNA Concentrations (cells/L), and COD Removal Efficiencies (%) During 15°C Operation.

SRT	High DO			Intermediate DO			Low DO		
	MLVSS	16S rDNA	COD Eff.	MLVSS	16S rDNA	COD Eff.	MLVSS	16S rDNA	COD Eff.
20-Day	2293	3.7×10^{11}	87%	2838	9.8×10^{11}	86%	2974	16.0×10^{11}	88%
10-Day	1398	3.7×10^{11}	84%	1645	7.8×10^{11}	84%	1691	16.0×10^{11}	86%
5-Day	739	3.1×10^{11}	82%	838	8.0×10^{11}	81%	772	5.7×10^{11}	84%
2-Day	321	1.5×10^{11}	81%	395	4.3×10^{11}	81%	330	2.6×10^{11}	80%

Table ES-3. Average MLVSS (mg/L), Total Bacterial 16S rDNA Concentrations (cells/L), and COD Removal Efficiencies(%) During 10°C Operation.

SRT	High DO			Intermediate DO			Low DO		
	MLVSS	16S rDNA	COD Eff.	MLVSS	16S rDNA	COD Eff.	MLVSS	16S rDNA	COD Eff.
20-Day	2939	18.0×10^{11}	88%	2571	5.0×10^{11}	78%	2252	16.0×10^{11}	80%
10-Day	1698	10.0×10^{11}	85%	1398	10.0×10^{11}	75%	1247	9.5×10^{11}	78%
5-Day	908	4.8×10^{11}	83%	724	4.7×10^{11}	73%	556	6.9×10^{11}	74%
2-Day	391	3.7×10^{11}	82%	368	3.2×10^{11}	70%	276	3.3×10^{11}	70%

20°C operation. However, at this temperature the DO level did not appear to have a significant effect on the *N. oligotropha*-type AOB population, with the exception of the 2-day SRT reactor. Although the extent of nitrification is clearly reduced in all four SRT reactors at the lowest DO level, this trend was not reflected in the *N. oligotropha*-type AOB concentration in the 20-, 10-, and 5-day SRT reactors.

Comparison of Tables ES-4 and ES-5 revealed that average *N. oligotropha*-type AOB concentrations generally decreased when the temperature was lowered from 20°C to 15°C. At 15°C, the *N. oligotropha*-type AOB concentration varied little with SRT and DO, except for the marked decrease in the 2-day SRT reactor at the lowest DO level. The AOB concentration drop in this system corresponded to a large drop in ammonia oxidation efficiency. Ammonia treatment performance in the 20- and 10-day SRT systems remained fairly stable at 15°C, whereas

Table ES-4. Average Influent and Effluent Ammonia Concentrations (mg N/L) and *N. oligotropha*-type *amo*A Gene (i.e., AOB) Concentrations (cells/L) Measured During 20°C Operation.

SRT	High DO			Intermediate DO			Low DO		
	Influent	Effluent	AOB	Influent	Effluent	AOB	Influent	Effluent	AOB
20-Day	18	0.2	39.0×10^8	20	<0.1	13.0×10^8	20	1.0	24.0×10^8
10-Day	18	0.1	18.0×10^8	20	<0.1	5.0×10^8	20	3.0	7.4×10^8
5-Day	18	0.1	2.5×10^8	20	<0.1	2.5×10^8	20	7.0	5.7×10^8
2-Day	18	6.0	7.6×10^8	20	11.0	3.7×10^8	20	21	1.4×10^8

Table ES-5. Average Influent and Effluent Ammonia Concentrations (mg N/L) and *N. oligotropha*-type *amo*A Gene (i.e., AOB) Concentrations (cells/L) Measured During 15°C Operation.

SRT	High DO			Intermediate DO			Low DO		
	Influent	Effluent	AOB	Influent	Effluent	AOB	Influent	Effluent	AOB
20-Day	17	0.2	6.9×10^8	24	<0.1	4.7×10^8	17	0.2	4.1×10^8
10-Day	17	0.4	4.1×10^8	24	0.2	1.8×10^8	17	3.0	4.6×10^8
5-Day	17	6.0	6.9×10^8	24	6.0	6.2×10^8	17	11.0	2.8×10^8
2-Day	17	12.0	4.1×10^8	24	17.0	3.2×10^8	17	16.0	0.14×10^8

Table ES-6. Average Influent and Effluent Ammonia Concentrations (mg N/L) and *N. oligotropha*-type *amo*A Gene (i.e., AOB) Concentrations (cells/L) Measured During 10°C Operation.

SRT	High DO			Intermediate DO			Low DO		
	Influent	Effluent	AOB	Influent	Effluent	AOB	Influent	Effluent	AOB
20-Day	25	0.2	4.1×10^8	23	<0.1	1.2×10^8	21	0.1	0.78×10^8
10-Day	25	6.0	0.91×10^8	23	10.0	1.9×10^8	21	6.0	25.0×10^8
5-Day	25	14.0	0.39×10^8	23	15.0	0.78×10^8	21	17.0	1.8×10^8
2-Day	25	21.0	0.013×10^8	23	21.0	0.023×10^8	21	20.0	0.025×10^8

efficiency decreased incrementally in the 5- and 2-day SRT reactors with each shift in DO, culminating in complete nitrification failure in the 2-day SRT reactor at the low DO level. The trends in ammonia oxidation were only mirrored by the *N. oligotropha*-type AOB concentrations in the 5- and 2-day SRT reactors.

Tables ES-5 and ES-6 also show that the concentration of *N. oligotropha*-type AOB generally decreased with decreasing temperature. Additionally, the *N. oligotropha*-type AOB concentration tended to decrease with decreasing SRT during operation at 10°C. However, no clear correlation between AOB cell concentration and DO level was noted at this temperature. Interestingly, nitrification performance (based on chemical data) in the 10-day SRT reactor actually improved when the DO level was lowered. The *N. oligotropha*-type AOB levels obtained from real-time PCR reflect this trend. The 2-day SRT system was at complete nitrification failure throughout the course of the 10°C treatment. This condition was reflected in a fairly stable *N. oligotropha*-type AOB concentration of approximately 2 x 10^6 cells/L, which was over two orders of magnitude lower than the concentration measured when nitrification was occurring in this reactor (i.e., higher temperatures).

Attempts were made to compare the percentage of AOB measured in the reactors using real-time PCR and theoretical values estimated from design equations using typical kinetic parameters (Table ES-7). Measured percentages were computed by dividing the *N. oligotropha*-type AOB concentration (cells/L) by the total bacterial 16S concentration (cells/L). The theoretical values were calculated using the approach outlined in Rittmann and McCarty (2001). The values measured using PCR are much lower than the theoretically predicted values. The extremely low percentage of ammonia oxidizers measured in this study suggests that AOB other than *N. oligotropha*-type were dominant in the bench-scale system. Since multiple (at least 15 strains) AOB can exist (Purkhold et al., 2000), additional work is necessary to identify all of the major AOB species.

Table ES-7. Comparison of Percent AOB Based on Theoretical Calculations and Real-Time PCR Analysis.

	20°C		15°C		10°C	
SRT (days)	Theoretical	Measured	Theoretical	Measured	Theoretical	Measured
20	13.0%	0.3 (±0.4)%	9.7%	0.12 (±0.02)%	13.3%	0.02 (±0.01)%
10	9.8%	0.3 (±0.2)%	9.1%	0.1 (±0.04)%	10.3%	0.01 (±0.01)%
5	8.9%	0.07 (±0.09)%	8.3%	0.3 (±0.04)%	5.3%	0.01 (±0.001)%
2	6.2%	0.7 (±0.3)%	Unstable Nitrification	0.3 (±0.2)%	Unstable Nitrification	0.0004 (±0.0004)%

Nitrospira-type Nitrite-Oxidizing Bacteria

Real-time PCR measurements for total *Nitrospira spp.* 16S rDNA were made on samples collected from reactors operating at 20°C, 15°C, and 10°C. Since the number of *Nitrospira* 16S rDNA copies per cell is unknown, the assumption for this study was based on the average number of copies of 16S rDNA per cell in other nitrifying bacteria, such as *Nitrosomonas* and *Nitrobacter* (Navarro et al., 1992; Klappenbach et al., 2001). As shown in Table ES-8, the *Nitrospira* concentration remained fairly stable during operation at 20°C in the 20- and 10-day SRT reactors at each DO level and in the 5-day SRT reactor at the high and intermediate DO levels. The average *Nitrospira* concentration in the 5-day SRT reactor decreased by almost one-half at the low DO level. The *Nitrospira* concentration in the 2-day reactor decreased with decreasing DO and was significantly lower than levels measured in the other SRT reactors at this temperature.

At the high and intermediate DO levels, nitrite oxidation efficiencies were similar in the 20-, 10- and 5-day SRT reactors and decreased as the SRT decreased below five days. In the 2-day SRT system, nitrite efficiency decreased incrementally each time the DO level was lowered. These trends were mirrored in the *Nitrospira* cell densities. Additionally, nitrite oxidation decreased incrementally below an SRT of 10-days at the low DO level, which was also observed in the cell numbers of *Nitrospira*. Multiple linear regression analysis of the low DO level data

Table ES-8. Average Effluent Nitrite Concentration (mg N/L), *Nitrospira* 16S rDNA (i.e., NOB) Concentration (cells/L), and Nitrite Removal Efficiency (%) During 20°C Operation.

SRT	High DO			Intermediate DO			Low DO		
	Effluent	NOB	NO_2^- Eff.	Effluent	NOB	NO_2^- Eff.	Effluent	NOB	NO_2^- Eff.
20-Day	0.1	19.0 x 10^9	99%	<0.1	10.0 x 10^9	100%	0.1	13.0 x 10^9	99%
10-Day	0.1	29.0 x 10^9	99%	<0.1	23.0 x 10^9	100%	0.2	25.0 x 10^9	98%
5-Day	0.3	13.0 x 10^9	98%	0.3	13.0 x 10^9	99%	3.9	7.2 x 10^9	61%
2-Day	1.2	2.3 x 10^9	86%	1.1	0.53 x 10^9	80%	0.3	0.19 x 10^9	53%

showed that both SRT and available nitrite concentration exerted a significant impact on the nitrite-oxidizing bacteria (NOB) concentrations at a 95% confidence interval.

Comparison of Tables ES-8 and ES-9 shows that *Nitrospira* concentrations in the 20- and 10-day SRT reactors were not significantly affected by the temperature drop from 20°C to 15°C. The decrease in temperature did yield a decrease in *Nitrospira* levels in the 5-day SRT system; however, the effect of temperature on the 2-day SRT system was inconclusive. *Nitrospira* concentrations at 15°C appeared to decrease with decreasing SRT; however, levels actually increased in the 20-, 10- and 2-day SRT reactors when the DO levels were lowered. Based on chemical analysis, the 20- and 10-day SRT systems achieved ≥98% nitrite oxidation throughout the 15°C study. The 5- and 2-day SRT reactors showed incremental decreases in nitrification performance with decreasing DO levels. However, these trends were not reflected in the measured *Nitrospira* cell concentrations.

Table ES-9. Average Effluent Nitrite Concentration (mg N/L), *Nitrospira* 16S rDNA (i.e., NOB) Concentration (cells/L), and Nitrite Removal Efficiency (%) During 15°C Operation.

SRT	High DO			Intermediate DO			Low DO		
	Effluent	NOB	NO_2^- Eff.	Effluent	NOB	NO_2^- Eff.	Effluent	NOB	NO_2^- Eff.
20-Day	0.2	12.0 x 10^9	99%	0.1	31.0 x 10^9	99%	0.2	120.0 x 10^9	99%
10-Day	0.3	18.0 x 10^9	98%	0.3	26.0 x 10^9	98%	0.2	84.0 x 10^9	98%
5-Day	0.6	4.3 x 10^9	93%	0.9	4.5 x 10^9	92%	0.4	2.7 x 10^9	86%
2-Day	0.4	0.67 x 10^9	81%	0.6	0.81 x 10^9	78%	<0.1	1.8 x 10^9	73%

Tables ES-9 and ES-10 show that the concentration of *Nitrospira* in the 20- and 10-day SRT reactors was not significantly affected by the drop in temperature from 15°C to 10°C. However, average concentrations of *Nitrospira* tended to decrease with decreasing temperature for the 2- and 5-day SRT systems. Although average *Nitrospira* concentrations generally decreased with decreasing SRT at the 10°C operation, no clear correlation was observed between NOB concentration and DO level in the reactors at this temperature. Based on chemical analysis, the 20-day SRT reactor consistently achieved high levels of nitrification at all DO levels throughout the study; however, nitrification performance in the 10-day SRT reactor actually improved when the DO level was lowered. The nitrite oxidation efficiency in

Table ES-10. Average Effluent Nitrite Concentration (mg N/L), *Nitrospira* 16S rDNA (i.e., NOB) Concentration (cells/L), and Nitrite Removal Efficiency (%) During 10°C Operation.

SRT	High DO			Intermediate DO			Low DO		
	Effluent	NOB	NO_2^- Eff.	Effluent	NOB	NO_2^- Eff.	Effluent	NOB	NO_2^- Eff.
20-Day	0.2	97.0×10^9	99%	0.5	43.0×10^9	97%	0.1	53.0×10^9	100%
10-Day	2.4	20.0×10^9	87%	0.5	34.0×10^9	96%	0.3	25.0×10^9	98%
5-Day	0.6	1.7×10^9	92%	0.5	0.72×10^9	91%	0.3	1.2×10^9	88%
2-Day	0.1	0.41×10^9	90%	0.1	0.26×10^9	73%	<0.1	0.39×10^9	76%

the 5-day SRT reactor decreased significantly at the low DO level, while the 2-day SRT reactor was at complete nitrification failure throughout the course of 10°C operation. The trends in nitrite oxidation were reflected in the *Nitrospira* levels found in these reactors, with the exception of the 5-day SRT reactor.

The study also compared the percentages of NOB measured in the reactors using real-time PCR for *Nitrospira* 16S rDNA and theoretical values estimated from design equations using typical kinetic parameters (Table ES-11). Measured percentages were computed by dividing the *Nitrospira* concentration (cells/L) by the total bacterial 16S concentration (cells/L). The theoretical values were calculated using the approach outlined by Rittmann and McCarty (2001). The values measured using PCR are similar to theoretically predicted values and are in much better agreement than the AOB values, which suggests that *Nitrospira spp.* were the dominant nitrite oxidizers in the municipal bench-scale treatment systems. Additionally, a strong logarithmic relationship between the nitrite oxidation rate and NOB concentration was established.

Table ES-11. Comparison of Percent NOB Based on Theoretical Calculations and Real-Time Analysis.

	20°C		15°C		10°C	
SRT (days)	Theoretical	Measured	Theoretical	Measured	Theoretical	Measured
20	3.2%	2.2 (±1.2)%	4.0%	4.6 (±0.3)%	No steady-state data available	
10	3.2%	5.6 (±3.9)%	2.9%	4.8 (±1.5)%	2.4%	2.1 (±1.1)%
5	3.4%	3.3 (±1.0)%	2.3%	1.4 (±0.4)%	1.1%	0.2 (±0.007)%
2	1.2%	1.5 (±1.0)%	0.4%	0.6 (±0.5)%	Nitrification Failure	0.1 (±0.1)%

Industrial Wastewater Treatment

Total Bacterial 16S rDNA

Biosolids samples taken from the industrial bench-scale system were also used to quantify microbial populations. The average concentration of total bacteria in each SRT train ranged from approximately 1×10^{12} to 6×10^{12} cells/L during the four operating periods (Table ES-12). These levels are somewhat higher than those measured in the municipal bench-scale treatment system due to the higher influent COD feed concentration in the industrial system. The longer SRT reactors typically had a higher total bacteria number (per liter of mixed liquor) than the shorter SRT reactors. Total bacteria concentration increased linearly with increasing MLVSS concentration (in the longer SRT trains), indicating that higher numbers of bacteria were found in reactors containing higher MLVSS levels. Because of the relatively high influent COD concentrations, heterotrophic bacteria should predominate the microbial population in the industrial treatment system and, therefore, be closely aligned with the total bacteria concentration. When the total bacteria number was normalized to mass (mg) of MLVSS, the values obtained were essentially constant at approximately 1.3×10^9 cells/mg MLVSS in all four SRT trains during all operating periods. This suggests that SRT, DO level, influent ammonia concentration, and COD concentration do not significantly change the total bacteria number when normalized to unit mass of MLVSS.

Table ES-12. Average Total Bacterial 16S rDNA Concentrations (cells/L) in the Industrial Bench-Scale System During Each Operational Phase.

	Phase 1	Phase 2	Phase 3	Phase 4
	High DO & Low NH_4^+	High DO & High NH_4^+	Low DO& High NH_4^+	Low DO & Low NH_4^+
20-day	5.36 (±1.73) × 10^{12}	4.45 (±1.50) × 10^{12}	3.53 (±1.35) × 10^{12}	4.39 (±0.48) × 10^{12}
10-day	5.84 (±1.23) × 10^{12}	3.01 (±1.51) × 10^{12}	4.12 (±1.32) × 10^{12}	3.24 (±0.72) × 10^{12}
5-day	3.06 (±1.85) × 10^{12}	1.92 (±1.02) × 10^{12}	1.32 (±0.40) × 10^{12}	1.18 (±0.26) × 10^{12}
2-day	1.01 (±0.68) × 10^{12}	1.58 (±0.76) × 10^{12}	1.04 (±0.40) × 10^{12}	0.84 (±0.37) × 10^{12}

The somewhat narrow concentration range measured for total bacteria at each operating phase and in each system was likely the result of the high COD removal efficiency of these reactors. Average COD treatment performance was as good in the shorter SRT systems (2- and 5-day treatment trains) as in the longer SRT systems (10- and 20-day treatment trains), although total bacteria concentrations were lower in the shorter SRT reactors. However, COD treatment performance was not as consistent in the 2- and 5-day systems as in the 10- and 20-day systems. Longer SRTs (which yield higher total bacterial levels) seem to provide more stable operation.

Nitrosomonas nitrosa-type AOB

The *Nitrosomonas nitrosa*-type bacteria were expected to be the dominant AOB species in the industrial bench-scale system since this group was found to be the predominant AOB in the full-scale industrial WWTP. The real time-PCR assay used to quantify the *N. nitrosa*-type AOB in the full-scale system was similarly applied to quantify this group in the bench-scale system. In all four SRT trains, the number of *N. nitrosa*-type AOB were in the range of $10^5 \sim 10^6$ cells/L (Table ES-13). This AOB concentration range was very low when compared to the $10^9 \sim 10^{10}$ cells/L range measured in earlier full-scale industrial samples. Since the bench-scale reactors continued to effectively nitrify the incoming waste stream (especially in the longer SRT reactors), an AOB population shift must have occurred in this system. An assessment of relative population indicated that the number of *N. nitrosa*-type AOB was so low (<0.005% of the total bacteria measured) that this group of organisms was no longer responsible for the level of

Table ES-13. Average *N. nitrosa*-type Gene Concentrations (cells/L) in the Industrial Bench-Scale System During Each Operational Phase.

	Phase 1	Phase 2	Phase 3	Phase 4
	High DO & Low NH_4^+	High DO & High NH_4^+	Low DO& High NH_4^+	Low DO & Low NH_4^+
20-day	4.50 (±2.79) × 10^6	7.13 (±2.49) × 10^6	6.79 (±5.70) × 10^6	6.20 (±6.20) × 10^6
10-day	6.88 (±3.50) × 10^6	7.45 (±3.27) × 10^6	7.27 (±3.56) × 10^6	7.26 (±3.97) × 10^6
5-day	7.33 (±3.24) × 10^6	7.80 (±3.94) × 10^6	17.20 (±4.20) × 10^6	26.60 (±15.80) × 10^6
2-day	0.14 (±0.08) × 10^6	6.81 (±3.73) × 10^6	3.45 (±3.33) × 10^6	0.73 (±0.44) × 10^6

ammonia oxidation measured in the industrial bench-scale system. Additionally, the correlation between *N. nitrosa* concentration and ammonia-oxidation efficiency was not very strong.

Although the *N. nitrosa*-type bacteria were not the major AOB species in the industrial bench-scale system, the average concentration of this population increased in all four trains as influent ammonia concentrations were increased (during Phases 2 and 3). Additionally, the *N. nitrosa* population was not significantly affected by the low DO level during Phase 3 operation.

Nitrospira-type NOB

The *Nitrospira*-type bacteria were monitored as the dominant NOB in the industrial bench-scale system. Average concentrations of *Nitrospira*-type NOB in longer SRT trains (10- and 20-day) were higher than in shorter SRT trains (2- and 5-day) during all four phases (Table ES-14). During Phase 1 operation, when nitrite-oxidation did not significantly occur in the 2- and 5-day trains, an average *Nitrospira* concentration of approximately 1 × 10^9 cells/L was measured. This indicates that this type of NOB can survive low nitrite conditions without expressing significant nitrite oxidation activity. During the Phase 2 operation, the concentration of *Nitrospira*-type NOB increased in all four SRT trains, especially in the 5-, 10- and 20-day SRT systems. Based on chemical analysis, nitrite-oxidation performance remained high in the 10- and 20-day SRT trains, and significantly improved in the 2- and 5-day SRT trains. Growth of *Nitrospira*-type NOB occurred when additional nitrite became available after influent ammonia loading was augmented. During Phase 3 operation (DO = 0.5 mg/L), *Nitrospira* concentrations in the 2- and 5-day trains did not significantly change, while concentrations in the 10- and 20-day trains increased. Low DO operation did not significantly decrease the population of *Nitrospira*-type NOB.

Table ES-14. Average *Nitrospira* 16S rDNA Concentrations (cells/L) in the Industrial Bench-Scale System During Each Operational Phase.

	Phase 1	Phase 2	Phase 3	Phase 4
	High DO & Low NH_4^+	High DO & High NH_4^+	Low DO& High NH_4^+	Low DO & Low NH_4^+
20-day	2.59 (±1.68) × 10^{10}	11.10 (±9.06) × 10^{10}	46.70 (±32.40) × 10^{10}	16.50 (±2.02) × 10^{10}
10-day	2.22 (±0.70) × 10^{10}	5.22 (±4.00) × 10^{10}	16.90 (±10.40) × 10^{10}	10.30 (±2.16) × 10^{10}
5-day	0.12 (±0.06) × 10^{10}	2.11 (±1.36) × 10^{10}	2.08 (±1.91) × 10^{10}	3.56 (±1.86) × 10^{10}
2-day	0.10 (±0.07) × 10^{10}	0.18 (±0.03) × 10^{10}	0.17 (±0.17) × 10^{10}	0.38 (±0.09) × 10^{10}

In general, nitrite-oxidation performance remained fairly stable when the *Nitrospira*-type NOB concentration was higher than 1×10^{10} cells/L. Instability increased in nitrite oxidation performance when the NOB concentration dropped below this apparent threshold level.

ES-4 Recommendations for Future Research

This study developed and tested a promising new molecular method (real-time PCR) to quantify total bacteria and various groups of bacteria involved in nitrification. A number of different types of bacteria can be involved in the nitrification process, and the dynamics of the population can change over time. Future research must address identification of primary, if not all, groups of bacteria involved in important biological processes, such as nitrification. Specifically, additional research is needed in designing AOB primers and probes for the *amo*A gene or alternative targets.

The *amo*A gene is a definitive indicator for ammonia-oxidation. However, there is considerable DNA sequence variability in the *amo*A gene between ammonia-oxidizing species, making the design of primers and probes for all *amo*A genes extremely difficult. Therefore, multiple *amo*A gene assays will be needed to monitor all potential ammonia-oxidizing bacteria in biological samples.

As better, more efficient, and less expensive molecular procedures are developed, the possibility of a complete microbial assessment of various biological operations becomes a reality. Such information is required to move beyond "black box" approaches to waste treatment. With additional effort directed at incorporating newly developing automated instrumentation and more rapid thermocyclers, real-time PCR assays could be incorporated into routine wastewater testing in an analogous fashion to what is currently used in medical testing laboratories. The real-time PCR assay platform consists of DNA extraction followed by standardized PCR protocols. Once the platform is developed, target identification can be changed by modifying the primers and fluorescent probe used in the PCR assay. In this study, considerable effort was directed at designing the primers and probes used in the assays and developing the real-time PCR assay platform. However, once these methods were developed, students and technicians with moderate molecular biology laboratory skills were able to process hundreds of samples.

Although the insight provided by this study is not definitive, it provides a foundation for future work. Continued development of molecular methods for application to "dirty" waste systems is necessary if these methods are to be utilized by the wastewater industry. Additional research based on real-time PCR should be directed towards modification of the assays developed in this study for predicting nitrifier ***activity***, not just cell number. Whereas working with DNA has been traditionally easier than working with ribonucleic acid (RNA), recent improvements in the commercial kits available for RNA extractions and reverse transcriptase real-time PCR make RNA-based assays as straightforward as DNA-based assays. The main advantage of RNA-based PCR assays is that the level of particular messenger RNA (mRNA) transcript in a cell can be used to signal the general metabolic status or specific enzymatic activity depending on the mRNA target. Therefore, a logical extension of this study is to couple the primers and probes developed here and use them with RNA extracted from activated sludge samples in reverse transcriptase real-time PCR assays. Results from such an assay should provide a more rapid indication of problems that occur in the nitrifying population as changes in the *amo*A gene transcript are expressed in response to perturbations in operational or environmental conditions. Additional research should be directed at determining whether suitable primers and probes for RNA based assays can be designed to detect essentially all known ammonia-oxidizing bacteria.

The potential insight into the microbial dynamics of biological treatment systems that can be gained using a molecular approach is significant. Further development and application of this technology to problematic biological processes in waste treatment is warranted.

CHAPTER 1.0

INTRODUCTION

1.1 Background

Nitrification, the bacterially catalyzed oxidation of ammonia to nitrate, is an integral process in the global cycling of nitrogen and a vital component of most modern activated sludge wastewater treatment plants (WWTPs; Juretschko et al., 1998). In these plants, nitrification plays a key role in reducing the oxygen demand of ammonia and organic nitrogen. It is the preliminary step to complete removal of nitrogen compounds by denitrification (Hall and Murphy, 1980). Additionally, removal of ammonia from sewage is necessary, since this compound is toxic to aquatic life and can enhance eutrophication of receiving waters (Purkhold et al., 2000).

While the nitrification process has intensely researched over the past several decades, it is still well recognized as sensitive to distortions and consequently difficult to maintain in engineered systems. Many environmental and biological factors affect reaction dynamics (Marisili-Libelli and Giovannini, 1997), including the slow growth rates of nitrifying bacteria, their sensitivity to toxic shocks, and swings in temperature and pH.

A number of steady-state design and kinetic simulation models have been developed to describe nitrification in single-stage activated sludge systems (Bitton, 1999). Even so, the biological fundamentals behind most treatment technologies are often treated as "black boxes" with presumably known inputs and outputs (Keller et al., 2002). Nitrification is often viewed as more of a chemical than a biological process, and control systems are usually designed so that the pollutant concentrations in the effluent are maintained below prespecified levels with a minimum operational cost (Yuan and Blackall, 2002). Although the efficiency and robustness of a nitrifying WWTP mainly depend on the composition and activity of the microbial community, monitoring typically relies on chemical and physical parameters rather than microbiological data (Wagner et al., 2002) because suitable techniques to monitor and control the content of nitrifying bacteria in the biomass of a microbial community have been limited. Nitrifiers are difficult to cultivate, and traditional culture-based enumeration techniques result in an underestimation of both microbial number (Phillips et al., 2000; Ehrich et al., 1995) and diversity (Juretschko et al., 1998). Due to the problems associated with quantification of nitrifying bacteria using cultivation techniques, culture-independent molecular detection methods have recently been pursued, resulting in new tools that yield unprecedented opportunities for insight into the community structure of activated sludge. These tools are based on the detection and quantification of target molecules that are unique to specific microbial populations.

Together with traditional physical and chemical analyses, these techniques offer an opportunity to better define treatment performance and to improve process design and control.

1.2 Project Objectives

A complete understanding of microbial dynamics in activated sludge treatment is currently unrealized due, in part, to quantitative techniques that rely on culture-dependent methods. The bias inherent with cultivation techniques may result in isolates that do not necessarily represent those dominant in the system evaluated. In-situ analysis of microorganisms functioning in wastewater reactors will provide a greater understanding of how particular groups of organisms, such as nitrifiers, are impacted by system fluctuations. This information is key for improved process control and process design as well as for the development of more efficient biotreatment strategies.

A reliable and reproducible method for quantifying nitrifiers would be valuable for evaluating correlations between microbial population numbers, the effects of different operational parameters on cell density, and population changes in time and space, and may lead to more effective models for nitrification in WWTPs. The overall goal of this research was to elucidate the number of specific nitrogen (N)-utilizing bacteria under varying operational and environmental conditions during activated sludge treatment using molecular-based technologies. A primary objective of this research was to evaluate a new molecular technique (real-time polymerase chain reactions [PCR]) as a tool to quantify nitrifying bacteria in activated sludge. Specific objectives to accomplish this include development of PCR assays for:

- The eubacterial 16S ribosomal deoxyribonucleic acid (rDNA) gene (a measure of total bacteria present),
- The *Nitrosomonas oligotropha*- and *Nitrosomonas nitrosa*-type *amo*A gene (a measure of ammonia-oxidizing bacteria present), and
- The 16S rDNA gene from *Nitrospira* (a measure of nitrite-oxidizing bacteria present).

This research also attempted to utilize these new PCR assays, in addition to traditional physical/chemical methods, to evaluate nitrification treatment performance in combined carbon/nitrogen activated sludge systems treating municipal or industrial wastewater.

Ammonia removal efficiency and nitrification rates are typically controlled by the process conditions prevailing in the reactor and environmental factors. Important parameters include the solids retention time (SRT), hydraulic retention time (HRT), substrate concentration, pH, temperature, and dissolved oxygen (DO) concentration (Wilderer et al., 2002). In this research study, operational parameters were manipulated in order to stress the system and allow the evaluation of the real-time PCR assays over a range of nitrification performance efficiencies. Temperature, SRT and DO concentration were controlled (during the municipal portion of this work) to provide treatment efficiencies ranging from 0% to 100% nitrification. City wastewater was used as feed for the experimental system, so no control was exerted over ammonium concentration and pH; however, these parameters were monitored. For the industrial treatment component of this project, SRT, DO, and pH and were controlled, while influent ammonia levels were monitored and, during some phases of this research, supplemented. Industrial wastewater served as the feed during this part of the project.

CHAPTER 2.0

LITERATURE REVIEW

2.1 Introduction to the Activated Sludge Process

The activated sludge process utilizes a fluidized, mixed growth of microorganisms under aerobic conditions to consume the soluble and colloidal organic materials in wastewater as substrates. Thus it converts them to new biomass, carbon dioxide, and water through microbial respiration and synthesis (Reynolds and Richards, 1996). The carbon dioxide produced evolves as a gas and the biomass is removed by gravity sedimentation, leaving the wastewater free of the original organic matter. While the primary treatment objective in most activated sludge systems is the removal of soluble organic matter and oxidation of the carbon contained within, nitrification is also desired in many cases (Grady et al., 1999). In nitrifying activated sludge systems, the ammonia-nitrogen (NH_4^+-N) is converted to nitrate-nitrogen (NO_3^--N), which is then released in the effluent. If complete nitrogen removal is desired, it is then subjected to a denitrification step (Davis and Cornwell, 1998).

Activated sludge is a heterogeneous microbial culture composed mostly of bacteria, protozoa, rotifers, and fungi; however, the primary consumers of organic and inorganic wastes are the bacteria (Rittmann and McCarty, 2001). Like all organisms, bacteria derive energy and reducing power from oxidation reactions, which involve the removal of electrons. Two important sources of the electrons in activated sludge are organic and inorganic compounds that are present in the wastewater or released during treatment. Since removal and stabilization of organic matter are the most important processes of activated sludge treatment, bacteria that use organic compounds as their electron donor and their source of carbon for cell synthesis (heterotrophic) predominate in these systems (Grady et al., 1999). Although heterotrophic bacteria are quite diverse in activated sludge, they have certain common characteristics that are important for design and analysis. In general, heterotrophic bacteria in activated sludge are reasonably fast growers and can drive substrate concentrations quite low (Rittmann and McCarty, 2001). Bacteria that use inorganic compounds as their electron donor and carbon dioxide as their source of carbon in activated sludge (chemoautotrophic) are responsible for nitrification when ammonia-N and nitrite-N are used as electron donors (Rittmann and McCarty, 2001).

2.2 Fundamentals of Nitrification

Nitrification is the bacterially catalyzed oxidation of ammonia to nitrate, and is a vital component of most modern WWTPs. Large amounts of ammonia, produced through decomposition of organic material, are received by WWTPs for treatment. Nitrification plays a key role in reducing the oxygen demand of ammonia and organic nitrogen and is the preliminary step in the complete removal of nitrogen compounds by denitrification (Hall and Murphy, 1980). Lowering sewage ammonia levels is necessary since this nitrogen compound is toxic to aquatic life, can cause local oxygen depletion and eutrophication in receiving waters, and can affect chlorine disinfection efficiency (Cheremisinoff, 1996).

2.2.1 Nitrification Stoichiometry

The microbial oxidation of ammonia and ammonium ions to nitrate is a two-step process that occurs primarily through the coordination of two distinct chemoautotrophic groups of bacteria: ammonia-oxidizing bacteria (AOB) and nitrite-oxidizing bacteria (NOB). The chemoautotrophic organisms utilize ammonia or nitrite as an energy source, oxygen as the terminal electron acceptor, ammonia as the nitrogen source, and carbon dioxide as a carbon source. In the first step of nitrification, the AOB convert ammonia (NH_3) to nitrite (NO_2^-) via a two-step enzymatic process encoded by the ammonia monooxygenase (*amo*) and hydroxylamine oxidoreductase (*hao*) genes. The AMO enzyme catalyzes the oxygenation of ammonia to hydroxylamine:

$$NH_4^+ + O_2 + H^+ + 2e^- \rightarrow NH_2OH + H_2O \qquad (2.2.1\text{-}1)$$

The two electrons required in this process are derived from the oxidation of hydroxylamine to nitrite by hydroxylamine oxidoreductase enzyme (HAO):

$$NH_2OH + H_2O \rightarrow NO_2^- + 5H^+ + 4e^- \qquad (2.2.1\text{-}2)$$

Following this step, the NOB oxidize nitrite to nitrate in a single step catalyzed by the enzyme nitrite oxidoreductase according to the following reaction (Gray, 1989).

$$NO_2^- + 0.5O_2 \rightarrow NO_3^- \qquad (2.2.1\text{-}3)$$

The overall rate of reaction may be represented by combining equations 2.2.1-1, 2.2.1-2, and 2.2.1-3.

$$NH_4^+ + 2O_2 \rightarrow NO_3^- + 2H^+ + H_2O \qquad (2.2.1\text{-}4)$$

The overall nitrification reaction shows that the oxidation of ammonia to nitrate requires a high input of oxygen, about 4.5 kg for each kilogram of ammonia-N oxidized. Assimilative reactions are also occurring during nitrification according to equation 2.2.1-5 in which $C_5H_7NO_2$ is the empirical formula of a bacterial cell (Cheremisinoff, 1996).

$$NH_4^+ + 4CO_2 + HCO_3^- + H_2O \rightarrow C_5H_7NO_2 + 5O_2 \qquad (2.2.1\text{-}5)$$

By combining equations 2.2.1-4 and 2.2.1-5, the overall oxidation and assimilation reaction for ammonia is:

$$22NH_4^+ + 37O_2 + 4CO_2 + HCO_3^- + H_2O \rightarrow C_5H_7NO_2 + 21NO_3^- + 20H_2O + 42H^+ \qquad (2.2.1\text{-}6)$$

Equation 2.2.1-6 shows that alkalinity is destroyed during the nitrification process. The theoretical amount of alkalinity destroyed per milligram of ammonia-N oxidized is 7.14 mg/L (as $CaCO_3$). However, theoretical alkalinity destruction is rarely observed in data obtained from laboratory, pilot plant, or full-scaled studies. In most cases, the actual alkalinity destroyed is less than the theoretical value because mineralization of organic nitrogen to ammonia occurs in the activated sludge process and imparts alkalinity to the wastewater. The theoretical value is only valid for wastes in which no ammonia-N is obtained from mineralization of organic nitrogen, i.e., the total Kjeldahl nitrogen (TKN): ammonia ratio is one (Cheremisinoff, 1996).

2.2.2 Nitrifying Bacteria

According to many older textbooks, the model organism for ammonia oxidation is *Nitrosomonas europaea*. However, recent analyses of nitrifying activated sludge and biofilms showed that other ammonia-oxidizers are also important (Wagner et al., 2002). Based on ultrastructural properties, AOB had been divided into five genera, *Nitrosomonas*, *Nitrosococcus*, *Nitrosospira*, *Nitrosovibrio*, and *Nitrosolobus* (Madigan et al., 1997), but 16S rDNA analysis has caused the number of genera of the characterized bacteria to be reduced to three (Schramm et al., 1998; Wagner et al., 1998). *Nitrosococcus mobilis* has now been moved into *Nitrosomonas,* while *Nitrosospira*, *Nitrosovibrio*, and *Nitrosolobus* have all been reclassified as *Nitrosospira*, all of which fall within the beta subclass of the *Proteobacteria* (Schramm et al., 1998). The remaining members of the genus *Nitrosococcus*, which are all marine organisms, constitute a coherent group within the gamma subclass of the *Proteobacteria* (Wagner et al., 1998). Almost all recognized lineages of betaproteobacterial ammonia-oxidizers can be found in WWTPs (Wagner et al., 2002). While some nitrifying WWTPs are dominated by a single ammonia-oxidizing species (Juretschko, 1998), other plants harbor at least five co-existing ammonia-oxidizing populations that are present in significant numbers (Daims 2001a).

The NOB belong to several phylogenetic groups including *Nitrobacter* (alpha subclass *Proteobacteria*), *Nitrococcus* (gamma subclass *Proteobacteria*), *Nitrospina* (delta subclass *Proteobacteria*) and *Nitrospira*, which constitutes an independent phylum in the domain *Bacteria* (Belser, 1979; Juretschko et al.,1998). Known members of the genera *Nitrococcus* and *Nitrospina* are marine organisms, whereas both freshwater and marine organisms are found within *Nitrobacter* and *Nitrospira* genera (Madigan et al., 1997). Until the late 1990s, it was generally assumed that *Nitrobacter* were the key NOB in wastewater treatment (Bitton, 1999; Mobarry et al., 1996). However, several recent molecular studies demonstrated that *Nitrospira* occurred more frequently and at higher concentrations than *Nitrobacter* in nitrifying reactors (Burrell et al., 1998; Hovanec et al., 1998; Okabe et al., 1999; Schramm et al., 1998, 1999), suggesting that *Nitrospira* plays a larger role in nitrite oxidation than *Nitrobacter* in wastewater treatment systems. It has been postulated that the predominance of *Nitrospir*a-like bacteria over *Nitrobacter* in most WWTPs is a reflection of their different survival strategies. While *Nitrospira*-like nitrite-oxidizers are, according to data extracted from microelectrode fluorescence in situ hybridization (FISH) analyses, K-strategists and thus are well adapted to low nitrite and oxygen concentrations, *Nitrobacter* is postulated to be a relatively fast-growing r-strategist with low affinities to nitrite and oxygen (Schramm et al., 1999). Since nitrite-concentrations in most WWTP reactors are low, *Nitrospira* will likely outcompete *Nitrobacter* in these systems (Wagner et al., 2002).

Nitrifying bacteria have several unique growth characteristics that are important to their impact on and survival in biochemical operations. The first is that the maximum specific growth rates for both ammonia- and nitrite-oxidizers are low (compared to those of heterotrophs) for pure and mixed-culture systems under ideal conditions (Prosser, 1989). This is due to a low energy yield, which is linked to the oxidation of ammonium to nitrate (Henze et al., 1995). Consequently, if suspended growth reactors are operated in a manner that requires the bacteria to grow rapidly, the nitrifying bacteria will be lost from the system and nitrification will cease even though organic substrate removal continues. Second, the amount of biomass formed per unit of nitrogen is small. As a result, nitrifying bacteria may make a negligible contribution to the MLSS concentration even when they have a significant effect on process performance (Grady et al., 1999).

Under steady-state conditions, experimental evidence has shown that nitrite accumulation is insignificant (U.S. EPA, 1975; Wong-Chong and Loer, 1975; Charley et al., 1980), suggesting that ammonia oxidation is the rate-limiting step in nitrification. The lack of nitrite build-up can be attributed to the generally considerably larger maximum growth rate of nitrite-oxidizers compared to the maximum growth rate of ammonia-oxidizers and that both groups of bacteria have K_s values that are typically less than 1 mg/L (U.S. EPA, 1975). Nitrification is, therefore, typically modeled on the specific growth rate of ammonia-oxidizers.

2.3 Impacts of SRT on Activated Sludge Treatment Performance

The SRT is commonly used in activated sludge systems to control efficiency of wastewater treatment as well as the physical and biological characteristics of the sludge (Rittmann and McCarty, 2001). SRT, symbolized by θ_x, is defined as the average length of time a unit of biomass remains in the treatment system, or

$$\theta_X \equiv \frac{X_T}{\left(\dfrac{\Delta X_T}{\Delta t}\right)} \qquad (2.3\text{-}1)$$

where:

X_T = total active biomass in treatment system (kg)

$\left(\dfrac{\Delta X_T}{\Delta t}\right)$ = total quantity of active biomass withdrawn from the system daily (kg/day)

By definition, the appropriate value of SRT can be maintained by a controlled solids wasting policy in which a fraction, equal to the total system biomass or mixed liquor volatile suspended solids (MLVSS) over the SRT, is removed per day (Lawrence and Brown, 1976).

The SRT is used as a fundamental variable because it is functionally related to the steady-state specific growth rate of the biomass in a complete-mix reactor. Since at steady state, the mass of biomass wasted must equal the net mass formed; that is, the growth minus the loss due to lysis, death, and decay, it follows that the SRT is inversely proportional to the net average specific growth rate of the entire system (Benefield and Randall, 1980). Consequently, the SRT controls the concentration of the growth-limiting substrate in the reactor and determines important system characteristics, such as the electron acceptor requirement and the excess biomass production rate (Grady et al., 1999).

Garret and Sawyer (1951) were the first to directly apply the idea of SRT as a measure of activated sludge performance; however, the systematic use of this basic concept for the design of suspended growth biological treatment systems was not described until almost 20 years later by Lawrence and McCarty (1970). By incorporating the equations describing microbial growth into mass balance equations for biomass and substrate in the treatment system, Lawrence and McCarty were able to develop equation 2.3-2 which clearly shows the relationship between SRT and treatment performance. Detailed derivations for this equation and others described below can be found in the classic paper by Lawrence and McCarty (1970) and numerous textbooks, (e.g. Benefield and Randall, 1980; Grady et al., 1999; Rittmann and McCarty, 2001).

$$S = K\frac{1+b\theta_x}{\theta_x\left(Yq_{max} - b\right) - 1} \qquad (2.3\text{-}2)$$

where:

S = concentration of the rate-limiting substrate (mg/L)
K = half-saturation constant (mg/L)
b = endogenous decay coefficient (day^{-1})
θ_x = SRT (days)
Y = true yield (mg/mg)
q_{max}= maximum specific rate of substrate utilization (mg/ mg-d)

Equation 2.3-2 is also often written in terms of the maximum specific growth rate of the microorganisms, μ_{max}, which is defined by the following equation.

$$\mu_{max} = Yq_{max} \qquad (2.3\text{-}3)$$

Substitution of equation 2.3-3 into 2.3-2 yields:

$$S = K\frac{1+b\theta_x}{\theta_x\left(\mu_{max} - b\right) - 1} \qquad (2.3\text{-}4)$$

Lawrence and McCarty (1970) also described a minimum SRT value (θ_x^{min}) which represents the SRT below which complete failure of the biological process will occur. Below θ_x^{min}, organisms are removed from the treatment system at a rate greater than their maximum synthesis rate, so that eventually no organisms will remain in the reactor. θ_x^{min} is computed by setting S in equation 2.3-4 to the influent substrate concentration and solving for θ_x^{min}:

$$\theta_x^{min} = \frac{K + S^0}{S^0\left(\mu_{max} - b\right) - bK} \qquad (2.3\text{-}5)$$

where:

S^0 = influent concentration of the rate-limiting substrate (mg/L)

A relationship between the SRT and the steady-state mixed liquor microbial mass concentration was also developed by Lawrence and McCarty (1970) and is shown below in equation 2.3-6.

$$X = \frac{\theta_x}{\theta}\frac{Y\left(S^0 - S\right)}{1 + b\theta_x} \qquad (2.3\text{-}6)$$

where:

X = concentration of biomass in the reactor (mg/L)

θ = HRT (days)

Once the SRT for a system has been selected, the steady state biomass level can be determined. Although the mixed liquor suspended solids (MLSS) concentration range used in activated sludge systems is between 500 mg/L and 5000 mg/L, practical designs typically limit the MLSS concentration between 2000 mg/L and 5000 mg/L (Grady et al., 1999).

The equations described in this section can be applied to both the heterotrophic bacteria responsible for carbon oxidation and the autotrophic bacteria responsible for nitrification when the respective kinetic parameters and growth-limiting substrates for each population are considered. In the case of heterotrophic bacteria, the growth-limiting substrate is the biochemical oxygen demand (BOD), for the ammonia-oxidizers, it is ammonia-N, and for the nitrite oxidizers, it is nitrite-N.

2.3.1 Impacts of SRT on Heterotrophic Bacteria and Carbon Oxidation

The maximum specific growth rate, μ_{max}, is the kinetic parameter that exerts the most pronounced effect on SRT (Grady et al., 1999). Heterotrophic bacteria in activated sludge are characterized by relatively large maximum specific growth rates, resulting in low values of minimum SRT. A typical limiting value for the minimum SRT is 0.11 days for an activated sludge system operating at 20°C (Rittmann and McCarty, 2001). The range of SRTs necessary for removal of soluble organic matter from municipal wastewater is generally between 0.5 and 1.5 days (Grady et al., 1999). In practice, a safety factor is employed to account for uncertainty in the kinetic parameters, influent characteristics, and natural variability in the microbial community and guard against upsets in treatment performance. The SRT values typically used in full-scale WWTPs are in the range of 4 to 10 days for carbon oxidation (Metcalf and Eddy, 1991). The values represent conservative designs where high reliability in performance is required and have evolved from empirical practice over the years, rather than directly from treatment process theory (Rittmann and McCarty, 2001).

As discussed in the previous section, once the system SRT has been selected, the biomass concentration can be determined. The total biomass in the system is a function of the reactor volume and the biomass concentration in the reactor. Thus, for a fixed reactor volume, selection of the SRT determines the MLSS concentration in the system. In general, a higher SRT results in a greater amount of biomass in the system and therefore a higher MLSS concentration. Several studies have demonstrated that for a given waste stream and reactor concentration, MLSS concentration increases with SRT. Stover and Kincannon (1976) conducted a series of experiments on bench-scale activated sludge systems treating synthetic wastewater operated at 20°C and SRTs ranging from 2 days to 12 days. The resulting MLSS concentrations for an influent COD of 500 mg/L ranged from approximately 500 mg/L in the 2-day SRT reactor to 4000 mg/L in the 12-day SRT reactor. McClintock et al. (1993) obtained similar results when operating a pilot-scale activated sludge system to treat domestic wastewater operating at 20°C and SRTs of 1.2, 2.7, 5, and 15 days. The resulting MLVSS concentrations for influent COD values ranging from 257 mg/L to 344 mg/L were 424 mg/L, 658 mg/L, 1284 mg/L, and 1348 mg/L for the 1.2-, 2.7-, 5-, and 15-day SRT operation periods, respectively.

2.3.2 Impacts of SRT on Nitrification

The single most important factor that determines whether or not a WWTP will support nitrification is the SRT (Cheremisinoff, 1996). As discussed previously, for a complete-mix activated sludge process operating at steady state, the specific growth rate is equal to the inverse of the SRT. This implies that if a nitrifying culture is to be established, the reciprocal of the operating SRT must be less than the maximum specific growth rate (μ_{max}) of the ammonia-oxidizing population. If this requirement is not met, the rate of organism wasting from the system will be greater than the rate of production, so that washout of the nitrifying culture will soon occur (Benefield and Randall, 1980). Since the μ_{max} value for autotrophic nitrifying bacteria is very low, the corresponding minimum SRT may be quite high (see equation 2.3-5). Typical μ_{max} values for autotrophs are almost an order of magnitude lower than those of heterotrophs, suggesting a minimum SRT for nitrifying bacteria almost an order of magnitude larger (Grady et al., 1999). As a consequence, autotrophic organisms can be lost from bioreactors under conditions that allow heterotrophic bacteria to grow freely (Cheremisinoff, 1996). Therefore, special consideration must be given to the choice of the SRT in systems containing autotrophic bacteria, and it cannot be assumed that the conditions suitable for the removal of soluble organic matter are suitable for the conversion of ammonia-N to nitrate-N. The minimum SRT values reported for nitrification range from 2 to 20 days with the variations in reported values largely due to differences in temperature, pH, DO and system type (Cheremisinoff, 1996). Grady and coauthors (1999) suggest that a minimum safety factor of 1.5 be applied to the minimum SRT for the most slow-growing organisms required in the bioreactor to ensure that their washout does not occur, although larger safety factors may be required in some circumstances.

In addition to low μ_{max} values, nitrifying bacteria are also characterized by very low K_s values. Consequently, complete-mix activated sludge systems that contain autotrophic bacteria will have low ammonia-N concentrations even when the bacteria are growing relatively fast, resulting in very low concentrations whenever the SRT is large enough to ensure stable growth. This also means that the ammonia-N concentration will rise rapidly as the SRT is decreased to the point of washout (Grady et al., 1999). It is for this reason that nitrification has gained the reputation of being an all or none phenomenon, i.e., the extent of nitrification approaches 100% whenever the SRT is long enough to give stable growth and rapidly falls to zero as washout occurs.

Several studies have been conducted to determine the necessary SRT for effective nitrification. Strover and Kincannon (1976) compared one- versus two-stage nitrification in which the reactors operated under complete mix conditions at 20°C with a synthetic wastewater with an influent COD: NH_3-N ratio of 500:50 mg/L. The results of the study for the one-stage system showed that greater than 40% ammonia removal was achieved at an SRT of 2 days and that the extent of ammonia removal increased with SRT up to about 10 days then leveled off at 99%. Hanaki et al. (1990) obtained similar results when operating a laboratory-scale mixed-flow reactor treating a synthetic substrate containing 80 mg/L of ammonia-N as part of a study on nitrification at low DO levels. During operation at 25°C and sufficient DO, complete nitrification was observed in the reactor when the SRT was $\geq$3.8 days, while operation at an SRT of 1.5 days resulted in approximately 60% ammonia oxidation and over 40% nitrite oxidation. This finding indicated that this system could nitrify more than 30 mg/L of influent ammonia at an SRT of only 1.5 days. In a study comparing nitrification rates and efficiencies for a biological nutrient removal process and a conventional activated sludge process, McClintock et al. (1993) also noted that efficient nitrification could be achieved in reactors operating at low SRTs. In this experiment, a pilot-scale activated sludge system, treating domestic wastewater with influent ammonia-N levels ranging from 16-30 mg/L, was operated at SRTs of 15, 5, 2.7, and 1.5 days at

20°C. The researchers found that complete nitrification occurred at SRTs ≥2.7 days, while operation at an SRT of 1.5 days resulted in 79% nitrification.

2.4 Effects of Temperature on Treatment Performance

Temperature is an important consideration in the design of an activated sludge process because of the effect it has on the activity of wastewater microbial populations and the rate of biological reactions (Benefield and Randall, 1980). The biochemical reactions employed in microbial metabolism for substrate utilization and endogenous degradation are enzyme-catalyzed reactions, and reactions catalyzed by enzymes are temperature dependent (Reynolds and Richards, 1996). It is well known that bacterial growth is optimal at a certain temperature (Antoniou et al., 1990). The majority of biological treatment systems operate in the mesophilic temperature range, which is 10°C to 45°C (Gray, 1989). For mesophilic bacteria, the optimal temperature for bacterial growth occurs at approximately 35°C (Tortora et al., 1993). Below this temperature, the rate of enzymatically catalyzed biological reactions increases with temperature in a linear fashion. Above 35°C, the rate of activity will decline as enzyme denaturation occurs and will eventually reach zero at a temperature around 45°C (Benefield and Randall, 1980). Activated sludge systems operate at the temperature of the wastewater, which is generally between 12°C and 25°C, and thus below the optimum temperature for biological activity (Gray, 1989). Therefore, in activated sludge systems, removal efficiency tends to increase with temperature.

From the preceding discussion, it is evident that the physiological state of bacteria can be directly affected by temperature changes. Temperature might affect the population distribution as well, with a correspondingly unpredictable effect on kinetic and stoichiometric coefficients. Therefore, it is necessary to account for changes in the operating temperature of an activated sludge process and determine the effect such changes will have on the values of the biokinetic coefficients used in process design (Benefield and Randall, 1980). The temperature correlation, represented in equation 2.4-1, is a commonly used temperature adjustment technique for the kinetic and stoichiometric parameters characterizing activated sludge (Sayigh and Malina, 1978).

$$k_1 = k_2 \theta_t^{T-20} \qquad (2.4\text{-}1)$$

where:

T = temperature (°C)
k_1 = the unknown value of any parameter at temperature T
k_2 = the known value of any parameter at a reference temperature, usually 20°C
θ_t = temperature coefficient

The two parameters required to characterize biomass growth are μ_{max} and K. The maximum specific growth rate is a strong function of the prevailing temperature and is assumed to vary according to the above relationship over the range of normal operating temperatures (Antoniou et al., 1990). The half saturation constant describes how substrate concentration influences the specific growth rate, and thus the impact of temperature on it is less clear, with it increasing under some circumstances and decreasing under others (Grady et al., 1999).

Consequently there is no consensus about its relationships to temperature, and each situation must be experimentally determined.

As noted in equation 2.3-3, biomass growth and substrate utilization are proportional to each other with the yield (Y) the proportionality coefficient. Research indicates that the yield will not substantially vary according to equation 2.4-1 (Benefield and Randall, 1980). Furthermore, there is some disagreement as to what a particular response will be to variations in temperature. Muck and Grady (1974) found that yield increased from 10°C to 20°C but then decreased as the temperature was increased above 20°C. However, Benefield et al. (1975) and Sayigh and Malina (1976) found that the yield value was basically independent of temperature within the range of 4°C to 20°C. Due to the uncertainty associated with the impact of temperature on the yield coefficient, most engineers assume that this parameter is independent of temperature, thereby allowing the same temperature coefficient to be used for both growth and substrate utilization (Grady et al., 1999).

Because the factors contributing to decay of bacteria are the same as those contributing to growth, it is logical to expect temperature to have similar effects on the microbial decay coefficient (b) and μ_{max}. Benefield et al. (1975) found that b varied according the relationship depicted in equation 2.4-1 between 15°C and 25°C, but deviated appreciably at 32°C. However, as 32°C is outside the operating range of most activated sludge plants, it is reasonable to assume that variations in b with temperature can be estimated by equation 2.4-1.

2.4.1 Impacts of Temperature on Heterotrophic Bacteria and Carbon Oxidation

Several studies have been conducted to assess the impact of temperature on the aerobic growth of heterotrophs. Characklis and Gujer (1979) have reported an average value of 1.09 for the μ_{max} temperature coefficient (θ_t) for heterotrophs, while Eckenfelder (1989) reported that the temperature coefficient is usually from 1.03 to 1.09. Grady et al. (1999) suggested using a θ_t value of 1.08. The variation in reported values is due to the difference between various activated sludge systems, and this is dependent upon the specific wastewater of interest (Reynolds and Richards, 1996). Benefield et al. (1975) found that b varied according to the Arrhenius relationship between 15°C and 25°C, and that θ_t varied between 1.02 and 1.06. Reynolds and Richards (1996) reported a slightly higher range of 1.065 to 1.085 for θ_t. Grady et al. (1999) suggested an average value of 1.04 for use in equation 2.4-1 for the microbial decay coefficient for heterotrophs. Due to the relatively few studies available that report effects of temperature on the parameters K_s and Y, and the lack of consensus among them as to whether they increased or decreased with increasing temperature, a θ_t value of 1 is generally used for these parameters (Grady et al., 1999). Although a temperature correlation relationship can be used to approximate the variation in reaction rates with temperature, it is recommended that, when possible, all biokinetic constants be evaluated at the aeration tank temperature expected under both critical summer and winter conditions (Grady et al., 1999).

In terms of removal efficiency, higher temperatures give rise to higher rates of BOD removal up to 35°C to 40°C, after which bacterial cells become reduced in size and number, thus growing in a dispersed phase and resulting in turbid effluents (Gray, 1989). However, experimental evidence indicates the process of carbon oxidation by heterotrophic bacteria is fairly insensitive to temperature because of the high solids levels maintained in activated sludge systems (Benefield and Randall, 1980). In a set of bench-scale activated sludge experiments, Sayligh and Malina (1978) found that at SRTs lower than 2 days, the efficiency of soluble BOD removal deteriorated appreciably as the temperature decreased from 31°C to 4°C. However, at SRTs of 3 days or more, effluent soluble BOD concentrations were <10 mg/L over the temperature range studied, suggesting that BOD removal efficiency was relatively independent of temperature variations between 4°C and 31°C. McClintock et al. (1993) reported similar results with a pilot-scale activated sludge system treating domestic wastewater with an average influent chemical oxygen demand (COD) concentration of 110 mg/L to 344 mg/L. When operated at an SRT of 15 days, the system achieved COD removals of 88%, 91%, and 87% at 20°C, 15°C, and 10°C, respectively. At an SRT of 5 days, 93%, 96%, and 86% COD removal was observed at temperatures of 20°C, 15°C, and 10°C, respectively. This data suggests that variations in temperature had no significant effect on carbon treatment over the range of normal operating temperatures for activated sludge systems.

2.4.2 Impacts of Temperature on Nitrification

While nitrification has been found to occur over a range of temperatures, the overall rate of nitrification has been shown to decrease with decreasing temperature (Poduska and Andrews, 1975). Temperature effects on nitrification are primarily related to the slow growth rate of nitrifying bacteria, which is exacerbated at low temperatures (Ydstebo et al., 2000). As discussed previously, nitrifying bacteria have much lower μ_{max} values than heterotrophs at optimum conditions, and the literature suggests that nitrifiers are also more sensitive to changes in temperature than heterotrophs. This statement is evidenced by the larger temperature coefficients reported for nitrifiers than heterotrophs. For example, Characklis and Gujer (1979) reported an average θ_t of 1.11 for μ_{max} for nitrification, as opposed to 1.09 for heterotrophs. As with heterotrophic bacteria, the yield coefficient is generally assumed to be independent of temperature. This is due to the uncertainty associated with the impact of temperature on Y. However, unlike heterotrophs, for which temperature appears to have variable effects on K_s, increases in temperature cause the half-saturation coefficient for nitrifiers to increase. The most widely cited data is that of Knowles et al. (1965), who reported θ_t=1.13 for *Nitrosomonas* and 1.16 for *Nitrobacter*. In spite of the importance of temperature to nitrification, few studies have systematically studied the effects of temperature on the decay coefficient for nitrifying bacteria (Grady et al., 1999).

Nitrosomonas and *Nitrobacter* seem to be influenced similarly by temperature; however, there are conflicting reports about the relative effects of temperature on the maximum specific growth rates of these two genera of nitrifiers (Shammas, 1986). Characklis and Gujer (1979) reported average θ_t values of 1.12 and 1.07 for μ_{max} for *Nitrosomonas* and *Nitrobacter*, respectively. In contrast, Hall and Murphy (1980) reported θ_t values of 1.10 and 1.11 for *Nitrosomonas* and *Nitrobacter*, respectively. Randall and Buth (1984) also found *Nitrobacter* to have a larger temperature coefficient that *Nitrosomonas* over the temperature range of 5°C to 30°C. Their results indicated that although the maximum possible rate of nitrate formation may be substantially greater than the maximum rate of nitrite formation at 20°C, there is a lower critical temperature below which the rate of nitrite formation will exceed that of nitrate, resulting in a build-up of nitrite. In their experiments, the critical temperatures for a laboratory-scale and a full-scale reactor were 14°C and 12°C, respectively. Below the critical temperature, the rate of nitrate formation controls the overall nitrification rate.

Several studies have been conducted to assess the impact of temperature on the efficiency of nitrification. Wild et al. (1971) conducted temperature studies over the range of 5°C to 30°C, and found that nitrification occurred at all temperatures investigated. Their results indicated that an approximate straight-line relationship existed between nitrification rate and temperature. They found that the rate at 27°C was 90% of that at 30°C, and at 17°C was 50% of that at 30°C. They concluded that on the basis of temperature alone and the most adverse conditions considered possible, it may be necessary to increase the SRT by up to five times the SRT during the summer in order to accomplish complete nitrification in the colder seasons. In another study, McClintock et al. (1993) observed complete nitrification in a pilot-scale activated sludge system that was treating domestic wastewater at a 15-day SRT and temperatures of 10°, 15°, and 20°C, and at a 5-day SRT and temperatures of 10°C, 15°C, and 20°C. Partial nitrification was observed with a 5-day SRT during operation at 10°C, with 65% conversion of ammonia-N to nitrate-N. Their results showed that an SRT of between 5 and 8 days was needed to maintain complete nitrification at 10°C. Ydstebo et al. (2000) also demonstrated that nitrification could occur at low temperatures. They studied nutrient removal at low temperatures performed at a biological nutrient removal (BNR) plant in Norway. Results demonstrated effective nitrification at temperatures as low as 6°C can be accomplished by operating at higher SRTs to support nitrifier growth. Low temperatures impacts were minimized by increasing the SRT from 4 days at 15°C to 14 days at 6°C (Ydstebo et al., 2000). Thus, the literature suggests that while the rate of nitrification decreases with decreasing temperature, negative effects can be overcome (to a considerable degree) by increasing the SRT in the activated sludge system.

2.5 Effects of DO on Treatment Performance

2.5.1 Carbon Treatment

In the presence of an easily degradable substrate, heterotrophic bacteria can grow at low DO concentrations. Although it is undesirable for oxygen to be rate limiting in the removal of organic compounds, limited research has been performed to establish the oxygen half-saturation coefficient for mixed cultures of heterotrophs ($K_{O,H}$). A probable reason for the lack of work in this area is that population shifts in the microbial community, due to changes in DO concentration, make estimation of the value difficult (Grady et al., 1999).

The oxygen half-saturation coefficient (K_O) has been defined as the oxygen concentration where catabolic metabolism takes place at one half of the maximum rate. This half-saturation coefficient is an indicator of a microorganism's affinity for oxygen. Therefore, a low value of K_O indicates a high affinity for oxygen, and the ability of a microorganism to effectively utilize substrate even at low DO conditions. Because of the disparity in value for heterotrophs and autotrophs, K_O becomes especially important in combined carbon removal/nitrification activated sludge systems. Limited pure culture experimental data suggest that the heterotrophic oxygen half saturation coefficient is very low. Sinclair and Ryder (1975) conducted a study on the effects of DO on the behavior of *Candida utilis* grown in a chemostat. Using a glycerol medium, the $K_{O,H}$ for *C. utilis* was found to be approximately 0.08 mg/L. Similar results were obtained by Lau et al. (1984) when describing the growth kinetics of a floc former *(Citrobacter* sp.) and a filament *(Sphaerotilis natans*) obtained from activated sludge. The $K_{O,H}$ values for the floc former and filament were calculated to be 0.15 mg/L and 0.01 mg/L respectively. These low values indicate that a very low DO concentration would be required to affect carbon treatment performance, although it might influence competition between filamentous and floc forming bacteria. Henze et al. (1987) have adopted a standardized $K_{O,H}$ value of 0.2 mg/L for use in the Activated Sludge Model (ASM) No. 1.

Due to the low K_O value for heterotrophs, it is not surprising that several researchers have found that DO has little effect on carbon treatment performance. Chuang et al. (1997) conducted a study to determine the effects of DO on nutrient removal for concentrations ranging from 2.0 mg/L to 0.1 mg/L. A synthetic feed was introduced into the reactors at a COD concentration of 300 mg/L. The effluent CODs were 10 mg/L, 11 mg/L, and 11 mg/L for a 10-day SRT reactor operating at 2.0 mg/L, 0.5 mg/L, and 0.1 mg/L DO, respectively. Similar treatment efficiencies for other SRTs showed that DO had no impact on carbon treatment for any of the SRTs studied (5, 10 and 15 days). Munch et al. (2000) conducted a pilot study to establish the feasibility of upgrading a municipal WWTP to incorporate nitrification into a carbon treatment system. It was determined that effective carbon treatment could be accomplished when the reactor was set at a DO concentration of 0.5 mg/L. Lau et al. (1984) also found that effective carbon treatment could be accomplished at low DO in a chemostat treating synthetic wastewater. The reactor was operated at a 0.3 day SRT for DO concentrations of 6.1 mg/L, 0.35 mg/L, and 0.09 mg/L. Operation at these DO concentrations produced effluent COD concentrations of 21 mg/L, 40 mg/L, and 30 mg/L. These findings indicated that DO concentration had no effect on carbon treatment efficiency.

When treating a waste stream, it is important to maintain a sufficient microbial population for degradation of organic and inorganic compounds. Heterotrophic bacteria comprise the majority of the MLSS in activated sludge systems (Grady et al., 1999) and these organisms are capable of effective carbon treatment even at very low DO levels. Therefore, major changes in MLSS concentration would not be expected to occur at low DO unless the influent COD concentration varied. A number of studies have determined that MLSS concentration was more a function of influent organic carbon than DO. The study of DO and carbon removal by Lau et al. (1984) revealed no significant changes in MLSS for a chemostat treating a consistent synthetic waste stream at 6.1 mg/L, 0.35 mg/L, and 0.9 mg/L DO. The MLSS values were 502, 525, and 490 respectively for this experiment. Chuang et al. (1997) obtained similar results when treating a synthetic influent feed comprised of 300 mg/L COD at 2 mg/L, 0.5 mg/L, and 0.1 mg/L DO. Operating the system at a 5-day SRT, MLSS concentrations of 920, 1010, and 1030 were obtained for the three DO levels, respectively. Ng et al. (1989) conducted a study on biological treatment of a pharmaceutical wastewater using activated sludge. The reactor was operated at SRTs of 20, 6.67, 4, and 2.86 days with a DO concentration of approximately 0.3 mg/L. It was discovered that an increase in COD loading produced a corresponding rise in MLSS concentration. Therefore, the increase in biomass was attributed to the change in influent COD.

2.5.2 Nitrification

2.5.2.1 Ammonia Oxidation

The DO concentration necessary for non-DO limited nitrification has been established at 2 mg/L (Benefield and Randall, 1985; Tchobanoglous and Burton, 1991); however, this value can vary with SRT and organic loading. Recent findings have shown that oxygen requirements for nitrification are not the same for different SRTs. The DO concentration required for high SRTs is as little as 1 mg/L (Fillos et al., 1996; Stenstrom and Song, 1991). Dangcong et al. (2000) conducted a sequencing batch reactor (SBR) study using a high concentration of influent ammonia and discovered that a significant portion of the incoming ammonia was converted to nitrite even though DO levels in the rector were close to zero. The results indicate that the ammonia oxidizers adapted to the low DO. Additionally, Hanaki et al. (1990) found that ammonia oxidation could efficiently occur in a pure nitrification reactor at a DO level of 0.5 mg/L and a SRT of 3.8 days. However, the researchers found that ammonia oxidation was reduced when a significant organic load was introduced in the influent. Their results showed that ammonia oxidation was inhibited at low DO levels for all SRTs when 500 mg/L and 1000 mg/L COD was added. However, efficient ammonia removal was measured at a SRT >4 days when the organic substrate concentration was 160 mg/L. Hanaki et al. (1990) attributed the decrease in ammonia removal at high organic loadings to an increase in K. Higher ammonia concentrations were required at high organic loadings to maintain the growth rate at low DO (Hanaki et al., 1990). Hanaki et al. (1990) also studied the observed growth yield index (PY_{obs}) of AOB and NOB to determine the effects of DO on growth. The proportional biomass determination factor (P) represented the specific substrate utilization rate of AOB from a batch test in which harvested mixed liquor cells consumed the substrate. The observed growth yield (Y_{obs}) was calculated by the following equation:

$$Y_{obs} = \frac{Y}{1 + b\theta_c} \qquad (2.5.2.1\text{-}1)$$

The first part of this study was conducted in a pure nitrification environment that limited the amount of heterotrophs by restricting organic carbon levels. The results showed that PY_{obs} for AOB significantly increased at low DO levels while PY_{obs} for NOB did not change. Since the growth yield index increased, either P or Y_{obs} must have caused the increase. P can be influenced by heterotrophs. However, only a negligible amount was present, so P was considered to be constant. This suggested that an elevated Y_{obs} caused the increase in the observed growth yield index. However, the substrate utilization rate decreased because of a decrease in the maximum substrate utilization rate when low DO conditions occurred (Hanaki et al., 1990). Consequently, the high growth yield increased the amount of ammonia oxidizing biomass and compensated for the reduced ammonia oxidation rate per unit biomass.

2.5.2.2 Nitrite Oxidation

Ammonia oxidation has historically been considered the rate- limiting step in nitrification. However, more recent findings seem to indicate that at low DO levels, NOB are inhibited while AOB are relatively unaffected (Hanaki et al., 1990; Fillos et al., 1996; Dangcong et al., 2000). Nitrite oxidation is inhibited in a low DO environment due to a specific affinity for oxygen that is lower than that of the ammonia oxidizers (Laanbroek et al., 1994). For this reason, NOB have difficulty competing for the available oxygen and adapting to the environment. Bernet et al. (2001) proposed using DO to limit the amount of nitrate produced in nitrogen removal processes because nitrite is cheaper to convert to N_2 gas via denitrification. To determine the effects of DO on nitrifiers, the researchers introduced two parameters (γ_1 and γ_2) to describe the percentage of maximum growth rate for ammonia and nitrite oxidizers. These two parameters can be defined by the following equations:

$$\gamma_1 = \mu_1^{'}/\mu_1 = \frac{DO}{K_{NH_4^+,O_2} + DO} \qquad (2.5.2.2\text{-}1)$$

$$\gamma_2 = \mu_2^{'}/\mu_2 = \frac{DO}{K_{NO_2^-,O_2} + DO} \qquad (2.5.2.2\text{-}2)$$

where:

μ_1 = maximum ammonia oxidizer growth rate at excess DO (day^{-1})
μ_1' = maximum ammonia oxidizer growth rate at low DO (day^{-1})
μ_2 = maximum nitrite oxidizer growth rate at excess DO (day^{-1})
μ_2' = maximum nitrite oxidizer growth rate at low DO (day^{-1})

Equation 2.5.2.2-1 can be divided by equation 2.5.2.2-2 to yield the variation of growth ratios between ammonia and nitrite oxidizers under different DO concentrations (Bernet et al., 2001):

$$\gamma = \gamma_1/\gamma_2 = \frac{K_{NO_2^-,O_2} + DO}{K_{NO_4^+,O_2} + DO} \qquad (2.5.2.2\text{-}3)$$

The researchers operated a nitrifying biofilm at 0.5 mg/L DO to determine whether nitrite accumulation could be sustained. The experiment was conducted over a 110-day period and nitrite accumulation occurred throughout. Results indicated that γ for the AOB was close to 1 (0.976) while γ for NOB decreased to 0.120. The findings indicated that nitrite oxidizers were only growing at 12% of their normal rate and consequently could not remove all of the nitrite. It was noted that once the DO was restored to 50% saturation, complete conversion of nitrite to nitrate once again occurred. This result shows that nitrite oxidizers were always present in the biofilm (Bernet et al., 2001).

A significant buildup of nitrite in the effluent is problematic because nitrite is toxic to aquatic life at concentrations as low as 0.5 mg/L (Balmelle et al., 1992). High nitrite levels can also inhibit growth of floc forming organisms, which could cause settling problems (Kappeler et al., 1994). Another side effect of excess nitrite is the production of N_2O gas. Zheng et al. (1994) used a 10-day SRT and DO concentrations of 0.1, 0.2, 0.5, 1.7, and 6.8 to examine the effect of DO on nitrous oxide formation. At a DO of 0.1 mg/L, approximately 5.4% of the nitrified nitrogen was converted to N_2O gas. The conversion rate increased to 7.0% at the 0.2 mg/L DO concentration and then declined back to 5.4% at the 0.5 mg/L DO concentration. The actual amount of N_2O produced at 0.5 mg/L DO was higher than at 0.2 mg/L, but more ammonification occurred, causing the percentage to drop. It appears that N_2O will be produced to some extent regardless of the DO level. However, conversion rates are highest at low DO levels. This condition correlates to the DO concentration where nitrite oxidation is inhibited suggesting a close relationship between N_2O production and nitrite buildup in wastewater (Zheng et al., 1994).

2.6 Introduction to Molecular Methods

The accumulation and diversity of a nitrifying microbial community in activated sludge play a crucial role in the performance of wastewater treatment systems (Kim and Ivanov, 2000). Until recently, suitable techniques to monitor and control the content of nitrifying bacteria in the biomass of a microbial community have been limited (Rittmann, 2002). Much of the knowledge about the physiology of nitrifying bacteria and requirements for their growth has come from pure culture studies (Grady and Filipe, 2000). However, it is known that only a small proportion of bacteria found in natural and engineered environments can be cultivated under standard laboratory conditions (Hermansson and Lindgren, 2001). Due to the low maximum growth rates and low biomass yields of nitrifiers, pure culture isolation is extremely difficult, resulting in an underestimation of both the number and diversity of organisms (Prosser, 1989). Estimation of nitrifier kinetic parameters in mixed cultures of heterotrophs and autotrophs is difficult because nitrifying bacteria are morphologically indistinct from heterotrophs (Hall and Murphy, 1980). Without a method of estimating the fraction of microorganisms that are autotrophic in a mixed culture sample, it is impossible to estimate parameters such as specific growth rate or specific substrate removal rate. Routine bacteriological enumeration by spread-plate or most-probable-number (MPN) count are not suitable for determining the mass of nitrifying organisms within samples of activated sludge because of the difficulty in obtaining nitrifier isolates (Mendum et al., 1999). Thus, ecological studies of nitrifiers have been severely limited by the lack of reliable and convenient techniques for estimating cell or biomass concentrations (Phillips et al., 2000).

Only recently have tools become available that make the quantification of individual microbial populations possible. These tools fall within the broad range of molecular signature techniques and are based on the detection and quantification of target molecules that are unique for specific microbial populations (Oerther et al., 1999). A common target molecule is the 16S rDNA gene, which codes for ribosomal ribonucleic acid (rRNA) molecules needed for protein synthesis (Wilderer et al., 2002). The 16S rRNA molecule has approximately 1500 bases that can be sequenced and contains both highly variable and conserved regions (Rittmann and McCarty, 2001). Analysis of the variable regions of the 16S rRNA molecule can identify stretches of 15 to 20 bases that are unique to a given organism or set of similar organisms. These regions can be used as targets for the identification and quantification of different microorganisms present in complex microbial communities such as those found in activated sludge (Rittmann, 2002). Another target molecule is a gene that codes for catabolic enzymes restricted to a strain or related group of strains. For the nitrifying bacteria, the gene for the ammonia-monooxygenase (AMO) enzyme, which catalyzes the critical first step of ammonia oxidation to hydroxylamine, is a good target for assessing a community's potential to perform nitrification (Rittmann, 2002).

Before nitrifying bacterial concentrations in a particular treatment facility can be quantified, key microbes in that plant or reactor must first be identified. This is accomplished by obtaining an inventory of the microorganisms present and establishing a 16S rRNA gene clone library (Wagner et al., 2002). The first step in the identification process is to extract DNA from the bacterial cells and construct a data bank consisting of specific genes, which can be used to identify bacterial cells independent of cultivation. Extracted 16S rDNA molecules are amplified enzymatically using the PCR, which will be described in detail in Section 2.6.1. Following amplification, the 16S sequences are then cloned and maintained in *Escherichia coli*, which serves as a host and further multiplies the DNA copies as part of its growth cycle. These *E. coli* cells effectively act as a gene library, an inventory of rRNA of the original microbial population (Wilderer et al., 2002). The DNA fragments can be sequenced to reveal the identity of the corresponding bacteria by comparative analysis with 16S rDNA sequences of other bacteria (Rittmann and McCarty, 2001). Currently more than 17,000 16S rDNA sequences are maintained in public databases (Klappenbach et al., 2001).

An environmental library provides information about the composition of the microbial community, but does not provide a quantitative estimate of microbial abundance (Wilderer et al., 2002). In order to determine the relative abundance of the respective bacterial populations, the information gained from the environmental library must be incorporated into quantitative molecular techniques such as real-time PCR. Real-time PCR combines amplification with the use of oligonucleotide hybridization probes. Oligonucleotide probing exploits the unique sequences in the rRNA that were identified from the 16S rRNA gene clone library, by creating small DNA molecules (the probes) that are perfectly complementary to the target sequence (Rittmann and McCarty, 2001). The probe binds to the target site though hydrogen bonding under the conditions of the hybridization assay (Rittmann, 2002). If the complementary sequence is not present, hybridization does not occur, and the probe is washed away from the target site (Biesterfield et al., 2001). By attaching a reporter molecule to the oligonucleotide probes, hybridization can be monitored. Real-time PCR allows hybridization to be monitored continuously throughout the experiment enabling quantification of the target organism.

2.6.1 Polymerase Chain Reaction

Real-time PCR is a variation of the traditional PCR method. PCR is a relatively simple technique by which a DNA template is amplified many thousand or million-fold quickly and reliably (Saiki et al., 1988). The PCR process generates sufficient material for subsequent experimental analyses.

PCR is closely patterned after the natural principle of DNA replication. When any cell divides, enzymes (polymerases) make a copy of the DNA in each chromosome (Tortora et al., 1995). The first step in this process is to "unzip" the two DNA chains of the double helix. As the two strands separate, DNA polymerase makes a copy using each strand as a template. In addition to the DNA polymerase, two other components are required for DNA replication: a supply of the four nucleotide bases and a primer. The four nucleotide bases, adenine (A), cytosine (C), guanine (G), and thiamine (T) are the building blocks of DNA. The A on one DNA strand always pairs with the T on the other strand, whereas C always pairs with G. Thus, the two strands are said to be complementary to each other. DNA polymerases cannot copy a chain of DNA without a short sequence of nucleotides to "prime" the process (Tortora et al., 1995). Therefore, the cell has another enzyme (primase) that actually makes the first few nucleotides of the copy. This stretch of DNA is referred to as the primer (Saiki et al., 1985). Once the primer is made, the polymerase can continue to develop the new chain.

A typical polymerase chain reaction includes all the necessary components for DNA duplication: the sample of target DNA, large quantities of the four nucleotides in the form of deoxynucleotide triphosphates (dNTPs), large quantities of two oligonucleotide primers, a thermostable DNA polymerase, reaction buffer, magnesium and optional additives (Erlich et al., 1991). In PCR, the goal is not to replicate the entire strand of DNA but to replicate a target sequence of approximately 100 to 600 base pairs that is unique to the organism. The oligonucleotide primers mark the ends of the target sequence. These primers are short, synthetic sequences of single-stranded DNA typically consisting of 20 to 30 bases that are specific for the target region of the organism (Saiki et al., 1985). Two primers are included in the PCR, one for each of the complementary single DNA strands. The components of the reaction are mixed and the reaction is placed in a thermal cycler, which is an automated instrument that takes the reaction through a series of different temperatures for varying amounts of time. This series of temperature and time adjustments is referred to as one cycle of amplification (Erlich et al., 1991).

The initial step in a cycle denatures the target DNA by heating it to 95°C or higher for 15 seconds to 2 minutes (Erlich et al., 1991). In the denaturation process, the two intertwined strands of DNA separate from one another, producing the necessary single-stranded DNA template for the thermostable polymerase (Holland et al., 1991). Since the hydrogen bonds linking the bases are weak, they break at high temperatures, whereas the bonds between deoxyribose and phosphates, which are stronger covalent bonds, remain intact (Erlich et al., 1991).

The primers cannot bind to the DNA strands at such a high temperature, so for the second step of the PCR cycle the vial is cooled to approximately 40°C to 60°C (Erlich et al., 1991). At this temperature, the oligonucleotide primers can form stable associations (anneal) with the separated target DNA strands and serve as primers for DNA synthesis by the *Taq* polymerase. This step lasts approximately 30 to 60 seconds.

The final step of the reaction is to make a complete copy of the templates. The synthesis of new DNA begins when the reaction temperature is raised to the optimum for the thermostable DNA polymerase, which is approximately 74°C. *Taq* DNA polymerase is a recombinant thermostable DNA polymerase from the organism *Thermus aquaticus* and, unlike normal polymerase enzymes, is active at high temperatures (Saiki et al., 1998). *Taq* DNA polymerase begins the synthesis process at the region marked by the primers. It synthesizes new double-stranded DNA molecules, both identical to the original double stranded target DNA region, by facilitating the binding and joining of the complementary nucleotides that are free in solution (dNTPs) (Holland et al., 1985). The length of the products generated during PCR is equal to the sum of the lengths of the two primers plus the distance in the target DNA between the primers (Erlich et al., 1991).

The three steps in the PCR (the separation of the strands, annealing the primer to the template, and the synthesis of new strands) take less than 2 minutes. Each step is carried out in the same vial. At the end of a cycle, each piece of DNA in the vial has been duplicated. The cycle can be repeated 30 or more times (Wilderer et al., 2002). Each newly synthesized DNA piece can act as a new template, so after 30 cycles, one million copies of a single piece of DNA can be produced. Taking into account the time it takes to change the temperature of the reaction vial, one million copies can be ready in about 3 hours (Saiki et al., 1998).

2.6.2 Fundamentals of Real-time PCR

The principle weakness of the original PCR method is that it is not quantitative. To overcome this shortcoming, variations of PCR, such as real-time PCR, have been developed to allow quantification of specific targets present in a complex sample. Real-time PCR requires the addition of dual-labeled fluorogenic probes (Lee et al., 1993; Bassler et al., 1995; Livak et al., 1995a,b) that anneal to a target sequence located between the two primer binding sites during the PCR. Several different fluorescent probes can be used in real-time PCR including the TaqMan probe utilized in this study (Harms et al., 2003). The TaqMan probe is a linear oligonucleotide complementary to a target nucleic acid sequence, with a reporter fluorophore attached to the 5' end and a quencher fluorophore attached to the 3' end (Heid et al., 1996). Modification of the probe with a 3' blocking phosphate prevents extension of the annealed probe during amplification (i.e., because the 3' end is blocked, it cannot act as a primer [Lie and Petropoulos, 1998]). When the probe is intact, fluorescent energy transfer occurs, and the fluorescent emission from the reporter dye, 6-carboxy-fluorescein (FAM), is absorbed by the quenching dye, 6-carboxy-tetramethyl-rhodamine (TAMRA) (Heid et al., 1996). During the extension phase of the PCR cycle, the TaqMan probe that is annealed to the target sequence is cleaved by the 5' to 3' exonuclease activity of the extending Taq polymerase, resulting in the separation of the reporter dye from the quencher dye (Lie and Petropoulos, 1998). When this occurs, the reporter dye emission is no longer transferred efficiently to the quenching dye, resulting in an increase of the reporter dye fluorescent emission spectra (Heid et al., 1996). This process occurs in every cycle and does not interfere with the exponential accumulation of PCR product (Giulietti et al., 2001).

The increase in fluorescence emission can be read by a sequence detector in "real-time," during the reaction, and is a direct consequence of target amplification during PCR (Giulietti et al., 2001). A computer software program calculates the change in fluorescence emission according to the following equation (Heid et al. 1996).

$$\Delta R_n = R_n^+ - R_n^- \qquad (2.6.2\text{-}1)$$

where:

R_n^+ = the fluorescence emission of the product at each time point
R_n^- = the fluorescence emission of the baseline

Thus, ΔR_n expresses the probe degradation during the PCR process (Giulietti et al., 2001). The computer software constructs amplification plots using the fluorescence emission data that are collected during PCR amplification, in which the ΔR_n values are plotted versus cycle number. Typically, an amplification plot has a sigmoidal shape (Lie and Petropoulos, 1998). Within the initial flat part of the curve, the amount of reporter that has been cleaved is not large enough to elevate the emission intensity above the baseline (Giulietti et al., 2001). After a certain number of amplification cycles, enough probe cleavage has occurred so that the ΔR_n values exceed the baseline. The subsequent exponential phase in the curve represents a logarithmic amplification of PCR product. Finally, a plateau phase is reached and no significant further product accumulation takes place (Lie and Petropoulos, 1998). To delineate the beginning of the exponential phase, a threshold is chosen based on the variability of the baseline. Usually this is determined as 10 times the standard deviation of the baseline, set from cycles 3 to 15, although this can be manually changed for each individual experiment if necessary (Giulietti et al., 2001). The cycle at which the emission intensity of the sample rises above baseline is referred to as the threshold cycle (C_t) and is inversely proportional to the target sequence concentration (Heid et al., 1996). The higher the target concentration, the lower number of amplification cycles required to detect the rise in reporter emission above the baseline (Lie and Petropoulos, 1998). Therefore, this can be used as a quantitative measurement of the input target. The C_T of each sample is then compared to a standard curve and the result is a numerical value of the number of target sequences in the sample (Hall et al., 2002).

Real-time PCR has several advantages over other quantitative PCR methods that rely on end-point analysis. Because the determination of the target input copy number is performed during the exponential phase of the reaction, a more accurate estimation of sample concentration is obtained (Heid et al., 1996). In addition, the large dynamic range (5-9 logs) of real-time PCR assays permits the simultaneous analysis of samples with varying input concentrations without the worry that high copy number samples reach plateau while lower copy number samples continue to amplify exponentially (Lie and Petropoulos, 1998). Finally, real-time PCR is performed in a closed tube system and requires no post-PCR manipulation of samples, which reduces the risk of cross-contamination in the laboratory and also increases sample throughput (Heid et al., 1996).

2.7 Applications of Molecular Methods to Quantify Nitrifiers

Quantification of nitrifying bacteria has been attempted using several different methods, including:

- Quantitative blot hybridization (Hovanec and Delong, 1996; Mobarry et al., 1996; Rittmann et al., 1999),
- FISH (Wagner et al.1995, 1996; Mobarry et al. 1996; Schramm et al., 1999),
- Competitive PCR (Kowalchuck et al., 1999; Phillips et al., 2000; Stephen et al., 1999; Bjerrum et al., 2002), and
- Most recently, real-time PCR (Hermansson and Lindgren, 2001; Hall et al., 2002; this study-Harms et al., 2003).

2.7.1 Fluorescence In Situ Hybridization (FISH)

FISH has been used to successfully characterize nitrifying bacteria in wastewater treatment systems (Juretschko et al., 1998; Okabe et al., 1999; Schramm et al., 1996; Wagner et al., 1993 and 1995). Results to date have yielded qualitative information about community structure; for example, which organisms are present and where they are located with respect to other populations (Biesterfeld et al., 2001). The variability of microscopic images obtained for one sample and the presence of bacterial aggregates and non-transparent particles in environmental samples complicate the enumeration of cells by whole-cell FISH when analyzed by fluorescent microscopy and image analysis (Kim and Ivanov, 2000). This can be particularly problematic for nitrifying bacteria in activated sludge, which grow as microcolonies embedded in a flock of heterotrophic microorganisms (Wagner et al., 1998), making the heterogeneity of microscopic images high. Even so, FISH has been successfully applied for the quantification of nitrifiers in activated sludge in several studies (Wagner et al., 1996; Jurestschko et al., 1998; Daims et al., 2001). Wagner et al. (1996) and Jurestschko et al. (1998) performed studies on an intermittently aerated nitrification-denitrification basin of an industrial WWTP receiving wastewater from an animal waste processing facility. This particular WWTP receives sewage with exceptionally high ammonia concentrations (up to 5000 mg/L), resulting in high numbers of ammonia oxidizers compared to domestic sewage treatment plants. Wagner et al. (1996) reported ammonia-oxidizer cell densities of $6.6 \times 10^8 \pm 1.3 \times 10^8$ cells (13% of total cell counts) per mL of activated sludge based on FISH measurements in this industrial plant. The researchers used this value to estimate a specific activity per in-situ detected ammonia oxidizing cell of 0.00022 ± 0.000045 pmol/cell per hour. In a related study performed by Juretschko et al. (1998), FISH demonstrated *Nitrospira*-like bacteria comprised 9% of the total bacterial counts and ammonia oxidizers represented 10% to 20% of the total number of cells in this industrial treatment plant. Daims et al. (2001a) reported similar ammonia-oxidizer cell densities for a separate stage nitrifying municipal WWTP. The researchers reported that $9.8 \times 10^7 \pm 1.9 \times 10^7$ ammonia oxidizer cells/mL were found in the activated sludge, representing approximately 8.4% of total population measured by FISH. Based on the measured cell density, the average in-situ activity was calculated to be 2.3 fmol of ammonia converted to nitrite per ammonia oxidizer per hour.

2.7.2 Competitive PCR

Competitive PCR (cPCR) has previously been used to quantify AOB in environmental samples and has been demonstrated both with the 16S rDNA template (Mendum et al., 1999; Phillips et al., 2000; Bjerrum et al., 2002) and with the *amo*A gene (Kowalchuk et al., 1999; Mendum et al., 1999; Stephen et al., 1999; Daims et al., 2002b; Bjerrum et al., 2002; this study-

Dionisi et al., 2002a,b). However, the use of cPCR with this group of bacteria in activated sludge has been limited (Kowalchuk et al., 1999; Daims et al., 2001, Bjerrum et al., 2002, this study-Dionisi et al., 2002a,b). To date, the only studies of cPCR applications for NOB in activated sludge have been performed by researchers involved in this project (Dionisi et al., 2002a,b).

2.7.2.1 Ammonia Oxidizing Bacteria

Several researchers have suggested that the *amo*A gene, encoding the ammonia monooxygenase enzyme active site, may be a more reliable tool than 16S rDNA-based approaches for enumeration of AOB in environmental samples, since it reduces the number of non-target organisms detected (Aakra et al., 2001; Mendum et al., 1999; Bjerrum et al., 2002; Norton et al., 2002). The 16S rDNA approach for ecological investigations of AOB often yields an overestimation of both number and diversity of AOB, while *amo*A gives a better estimate of the actual number and activity of AOB, but might give less information concerning the distribution of the phylogenetic groups of AOB (Aakra et al., 2001). While it is likely that the *amo*A primers only amplify *amo*A genes from a subset of AOB present in a sample, this may be preferable to 16S rRNA primer based assays where non-AOB may contribute to the obtained population estimates (Bjerrum et al., 2002).

Kowalchuck et al. (1999) studied the distribution and community composition of beta subgroup proteobacterial ammonia oxidizers in different types of compost and composting materials with cPCR targeted at the *amo*A gene. In their experiments, a calf slurry-fed activated sludge sample was used as a positive control and was found to contain $1.7 \times 10^{12} \pm 1.1 \times 10^{12}$ cells/g of *Nitrosospira* cluster 3 ammonia oxidizers. Daims et al. (2001b) also measured the concentration of AOB by cPCR of an *amo*A sequence fragment, and used the results to calculate ammonia turnover rates in a nitrifying biofilm and two separate stage nitrifying activated sludge WWTPs, as well as a denitrifying biofilm as a negative control. Results were reported in average ammonia turnover rates per cell, assuming that 80% of the influent ammonia was oxidized to nitrite and 20% was assimilated by the bacteria. Cell numbers were not provided. The ammonia turnover rates were 0.048 pmol NH_4^+ per cell per hour for the nitrifying biofilm, and 0.016 and 0.043 pmol NH_4^+ per cell per hour for the two nitrifying activated sludge systems. The denitrifying biofilm had a much lower ammonia turnover rate of 0.003 pmol NH_4^+ per cell per hour. The researchers found that these results were similar to those obtained in pure culture studies, with turnover rates of 0.0018-0.023 pmol NH_4^+ per cell per hour as reported by Belser (1979). In another study, Bjerrum et al. (2002) used cPCR assays to estimate cell densities of AOB in many different environmental samples, such as intertidal mud flats, rice paddy soil, and activated sludge. Competitive PCR estimates of *amo*A gene copy number in activated sludge from Marselisborg Municipal WWTP aerobic activated sludge basin (Aarhus, Denmark) revealed an AOB density of approximately 10^8 cells/g wet weight. This cell concentration was in agreement with levels reported for other WWTPs (using FISH) including Wagner et al. (1996) who reported $6.6 \times 10^8 \pm 1.3 \times 10^8$ AOB cells per mL of activated sludge. In this study, Dionisi et al. (2002a) utilized cPCR (based on *amo*A gene fragments) to detect ammonia-oxidizers in a single sludge full-scale municipal WWTP and reported values of 4.3×10^4 ($\pm 0.6 \times 10^4$) to 4.1×10^5 ($\pm 0.3 \times 10^5$) *Nitrosomonas oligotropha*-type cells/mL activated sludge mixed liquor. While there appears to be a large discrepancy between these values and those obtained by other researchers, the difference can be attributed to the type of treatment process studied. Dionisi et al. (2002a) studied a single stage activated sludge process treating low nitrogen levels, whereas the other studies assessed separate-stage nitrification processes and/or systems with high influent nitrogen loading. It has been established that the fraction of biomass that is composed of nitrifiers is considerably larger in separate-stage nitrification when compared to that in the combined process (Benefield and Randall, 1980). It is also possible that the *amo*A gene fragment used in this study (Dionisi et al., 2002a) did not account for all of the AOB present in the activated sludge.

2.7.2.2 Nitrite Oxidizing Bacteria

Thus far, the only studies of cPCR applications for NOB in activated sludge have been performed by researchers involved in this project (Dionisi et al., 2002a,b). In their analyses of a full-scale municipal activated sludge treatment process, a PCR assay targeted toward *Nitrospira* 16S rDNA resulted in NOB concentrations ranging from 5.8×10^6 ($\pm 0.98 \times 10^6$) to 5.5×10^7 ($\pm 0.2 \times 10^7$) *Nitrospira* cells/mL of mixed liquor.

2.7.3 Real-time PCR

While real-time PCR has been used extensively in medical research, its application to environmental research has only recently been established (Suzuki et al., 2000; Bowers et al., 2000; Gruntzig et al., 2001; Hermansson and Lindgren, 2001; Becker et al., 2002; Yuan and Blackall, 2001; Hall et al., 2002; this study-Harms et al., 2003).

2.7.3.1 Ammonia Oxidizing Bacteria

Hermansson et al. (2001) used real-time PCR with primers and probes targeting a 16S rDNA region of AOB to quantify populations of AOB samples in arable soil. Cell numbers in both fertilized and unfertilized soil were studied to determine the effect of fertilization on soil AOB density. A two- to three-fold increase in AOB was found in the fertilized soil, suggesting that an increase in ammonia yields an increase in the number of AOB present. During this investigation, Harms et al. (2003) utilized real-time PCR for quantification of *N. oligotropha*-like AOB using an *amo*A assay in a municipal WWTP. Samples were collected and analyzed monthly for 1 year and were found to contain an average *N. oligotropha*-like AOB cell concentration of 7.5×10^6 cells/mL of mixed liquor.

2.7.3.2 Nitrite Oxidizing Bacteria

Yuan and Blackall (2002) used a real-time PCR assay targeting *Nitrospira* 16S rDNA to quantify NOB in lab-scale reactors treating municipal wastewater and found *Nitrospira* concentrations of approximately 3×10^6 cells/mL of mixed liquor. Hall et al. (2002) applied the same real-time PCR *Nitrospira* assay to samples from five full-scale municipal WWTPs. *Nitrospira* cell densities of 2.9×10^3, 4.1×10^7, 1.8×10^9, 1.2×10^4, and 2.3×10^{10} cells/mL of mixed liquor were reported for the WWTPs achieving 48.1%, 98.8%, 99.3%, 99.7%, and 99.8% nitrification, respectively. Ammonia in the influent for all plants ranged from 21 mg/L to 42 mg/L. Yuan and Blackall (2002) reported that, on the surface, nitrification performance appeared to be correlated with *Nitrospira* cell numbers for the plants investigated. This correlation was not upheld for one plant in which low *Nitrospira* numbers (1.2×10^4 cells/mL) were measured, but a high degree of nitrification (99.7%) occurred, suggesting that organisms other than *Nitrospira* were responsible for nitrite oxidation in this plant. Harms et al. (2003) reported cell densities of 1.7×10^7 to 1.2×10^8 *Nitrospira* cells/mL of mixed liquor when using a real-time PCR assay for *Nitrospira* 16S rDNA on 12 monthly samples from a full-scale municipal WWTP. These values are comparable to those measured by Yuan and Blackall (2002).

2.8 Attempts to Correlate Nitrifier Cell Densities with Process Performance

Thus far, few attempts have been made to correlate the number and type of nitrifying organisms in WWTPs with observed nitrification rates (Urbain et al., 1997; Rittmann et al., 1999; Biesterfeld et al., 2001; Biesterfeld and Figueroa, 2002) other than the determinations of specific activity rates in-situ as previously discussed (Wagner et al., 1996; Daims et al., 2001a,b). Urbain et al. (1997) presented the first example of an integrated analysis of activated sludge process performance using chemical, molecular, and mathematical tools. In the study, influent and effluent wastewater quality and sludge concentration were measured in a pilot-scale aerobic membrane bioreactor (MBR) operated at different SRTs in a municipal WWTP and compared with predicted values obtained using the multispecies mathematical model published by Furamai and Rittmann (1992). The accuracy of the model was demonstrated by comparing COD, NH_4-N and NO_3-N values at steady state. The proportion of nitrifiers and heterotrophs in the system were predicted by the model and compared to values measured using the slot-blot technique with oligonucleotide probes targeted towards the 16S rRNA. Urbain et al. (1997) found that there was good agreement between the model-predicted community structure and the slot-blot results. The percentages of nitrifiers predicted by the model were 7.1%, 8.9%, and 7.5% for SRTs of 5, 10, and 20 days, respectively, while the results of the molecular analysis showed that nitrifiers comprised 5.3%, 6.6%, and 12.1% of the total bacterial population for the 5-, 10- and 20-day SRTs, respectively.

Rittmann et al. (1999) also used the mathematical model of Furamai and Rittmann (1992) to link community structure with traditional measures of process performance for full-scale activated sludge WWTPs. In this study, normal operating results were obtained from activated sludge plants in France and the Netherlands and the mixed liquors from these processes were analyzed with oligonucleotide probes targeted to the 16S rRNA of ammonium oxidizers and all active bacteria. Slot-blot assays with oligonucleotide probes and FISH gave direct measurements of the ratio of ammonia-oxidizers to heterotrophs and were found to be consistent with the same ratio translated from the operating data using the mathematical model.

Biesterfeld et al. (2001) examined ammonia oxidation rates in a full-scale nitrifying trickling filter and found that a reasonable correlation existed between the ammonia-oxidizing populations quantified by FISH and observed ammonia removal. Correlation coefficients obtained from the application of a simple linear regression of the AOB FISH signal area and ammonia removal rate ranged from 0.72 to 0.98 for the six sample dates investigated. In a similar study, Biesterfeld and Figueroa (2002) showed that the percentage of AOB in bench-scale nitrifying trickling filters tracked the nitrate plus nitrite generation rates. The researchers defined tracking as having a correlation coefficient >0.500 between nitrate plus nitrite generation rates and percentages of AOB FISH signal area. The correlation coefficients calculated by simple linear regression between the two parameters ranged from 0.54 to 0.78 throughout their sampling campaign.

CHAPTER 3.0

MATERIALS AND METHODS

3.1 Municipal Bench-Scale Reactor System

3.1.1 Collection and Storage of Influent Wastewater

Influent wastewater for the municipal component of this study was obtained in batches from the Knoxville Utility Board's (KUB) Kuwahee WWTP, in Tennessee. Two steel 55-gallon (208.2 L) drums were filled with wastewater using a sump pump submerged in the primary clarifier overflow stream. These were transported to the laboratory. The volume of the two drums was sufficient to feed the experimental system for approximately 4 days. The wastewater drums were stored at 5°C in one of the laboratory's constant temperature rooms and allowed to cool overnight before transfer into the 55-gallon influent feed drum. At an influent flow rate of 19 mL/min, the feed drum was emptied approximately every 2 days. This necessitated storage of the second wastewater drum for 2 days in the constant temperature room. The stored wastewater was kept at 5°C and the drum was sealed in order to minimize any biological reactions during storage.

Because experimental temperatures were always maintained above 5°C, a method was developed to ensure that the influent wastewater remained cool during introduction to the system. The influent storage drum was fitted with a plate coil cooling sleeve (Tranter, Inc.) and a plastic-coated insulation jacket. A circulating chiller unit (Forma Scientific, Inc.) was used to cool a 50/50 blend of water and ethylene glycol to -20°C. This chilled solution was pumped in a continuous loop through the cooling sleeve and back to the chiller unit using an external peristaltic pump and insulated tubing. A small, submersible aquarium pump was placed at the bottom of the influent feed drum to gently circulate the stored wastewater in order to maintain homogenous temperature and solids content in the influent.

3.1.2 Experimental Treatment System

The experimental treatment system consisted of four bench-scale continuous stirred tank reactors (CSTRs) operating in parallel, each with an external secondary clarifier. Figure 3-1 depicts one of these treatment units, while Figure 3-2 shows the entire municipal treatment system. The system was designed to simulate a conventional activated sludge wastewater treatment process (Hawkins, 2000).

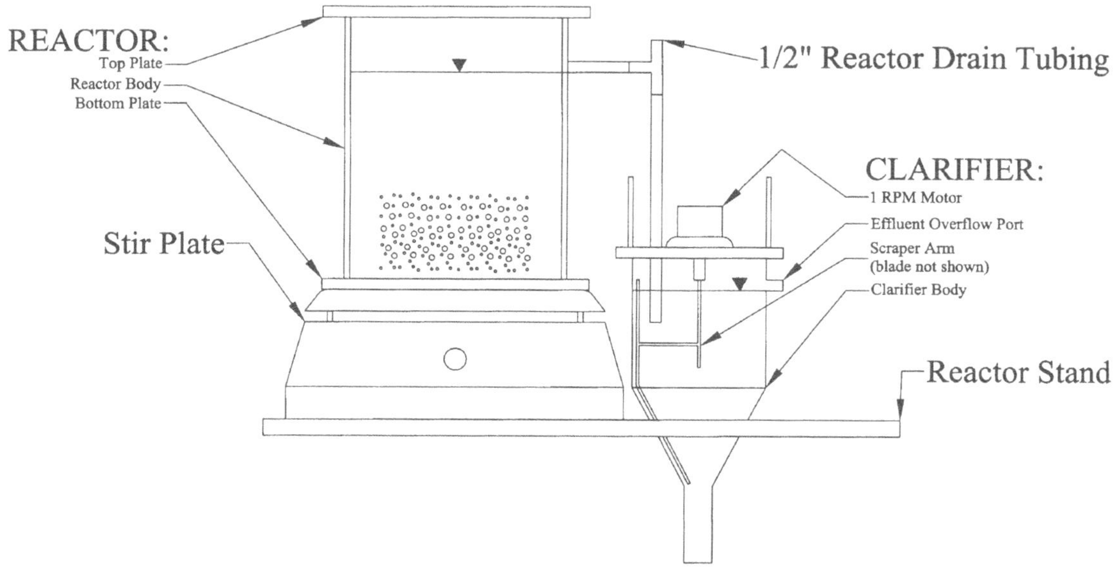

Figure 3-1. Diagram of One Treatment Unit.

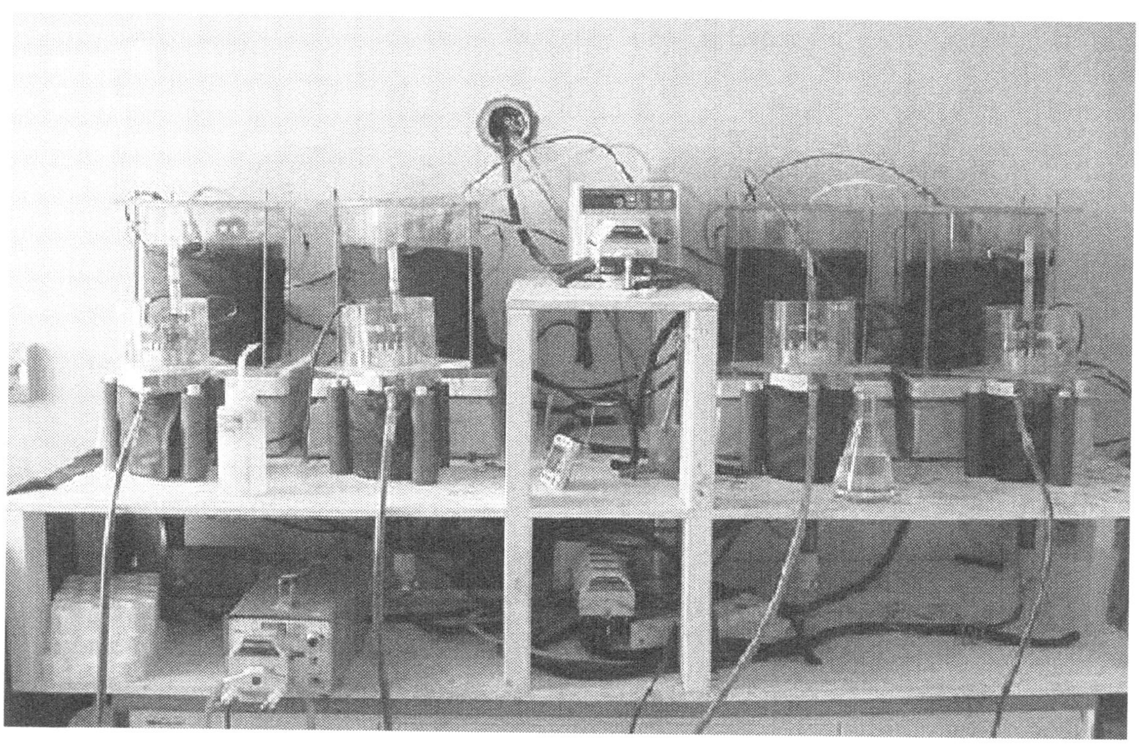

Figure 3-2. Treatment System Consisting of Reactors and Clarifiers.

Figure 3-3 shows a schematic diagram of the experimental treatment system. Influent wastewater was pumped from the feed drum into each of the four reactors using a continuous duty, standard drive, peristaltic pump (Cole Parmer, Inc.) and nylon tubing. The influent flow was split into four lines, each with its own pump head. In this manner, a single pump regulated each reactor's influent at a constant flow rate of 19 mL/min. A sample port was plumbed into the influent line of the 5-day SRT reactor to obtain influent wastewater grab samples.

Each CSTR was constructed of Plexiglas. The body was fabricated from a 12" (30.48 cm) section of 10" (25.4 cm) Plexiglas tubing and covered with 12" x 12" (30.48 cm x 30.48 cm) Plexiglas top and bottom plates, each etched to fit snuggly onto the body tube. The bottom plate was bonded onto the body to prevent leakage, however the top plate remained free to allow easy access to the reactor for sampling, wasting, cleaning, and maintenance. Each CSTR was placed on its own magnetic stir plate, which provided the necessary mixing energy to keep the contents of the reactor in suspension and completely mixed. The stir plates were each approximately 12" (30.48 cm) square and turned a 2" (5.08 cm) magnetic stir bar placed inside each reactor. When operated at one-half maximum turning speed, the stir plates provided a mixing power input of approximately 1500 kW/1000m^3, well above the required 14 kW/1000m^3 to ensure complete mixing in an activated sludge CSTR using mechanical mixing devices (Grady et al., 1999).

Each reactor was fitted with an overflow port to yield a total fluid volume of 10 L. The reactor mixed liquor, containing both treated wastewater and biological solids, flowed via gravity through the overflow port to the external secondary clarifier via drain tubing. The secondary clarifiers were constructed of blown glass and measured 6" (15.24 cm) in diameter at the top. The bottom was drawn into a conical section measuring 1.25" (3.18 cm) in diameter. To this bottom section a 3" (7.62 cm) section of 1.25" (3.18 cm) diameter tubing was welded, forming a reservoir for accumulated sludge. A recycle line of nylon tubing was affixed to bottom of each clarifier. An overflow port was cut into the side of each clarifier yielding a fluid volume of 2.9 L for each. A glass tube (*l* =1" [2.54 cm], *diameter*= 3/8" [0.95 cm]) was welded to each hole providing an overflow drain. Nylon tubing of appropriate length was fitted to the overflow drain lines and extended to four 40-L carboy containers placed on the floor below each of the clarifiers to collect effluent.

The clarifiers were each fitted with a 7" (17.78 cm) square Plexiglas top plate containing etched grooves similar to the CSTR top plates to ensure a snug fit. A 4" (10.16 cm) long section of 6" (15.24 cm) diameter Plexiglas tubing was bonded to the top of each top plate to form a protective housing. A scraper arm assembly was mounted on each top plate inside the protective housings to keep sludge from accumulating on the walls of the clarifiers. Each scraper arm assembly consisted of a 1-rpm motor mounted on the top plate and a scraping arm (connected to the motor via set screws through a hole cut into the top plate) that extended down into the clarifier body. The scraper arms were constructed from 1/8" (0.32 cm) diameter stainless steel wire and were formed to fit the shape of the clarifier walls. Automobile wiper blades were cut to the length of the scraper arms and attached to the arms such that a snug fit against the clarifier walls was obtained. The scraper motors were controlled with a timer to engage every 5 minutes and turn the blades approximately one-half turn.

The influent lines were attached to the CSTRs above the fluid level and extended toward the center of each reactor (approximately 1.5" [3.81 cm] from the wall) to avoid short-circuiting. The recycle lines were plumbed from the base of the clarifiers to the CSTRs and attached to the

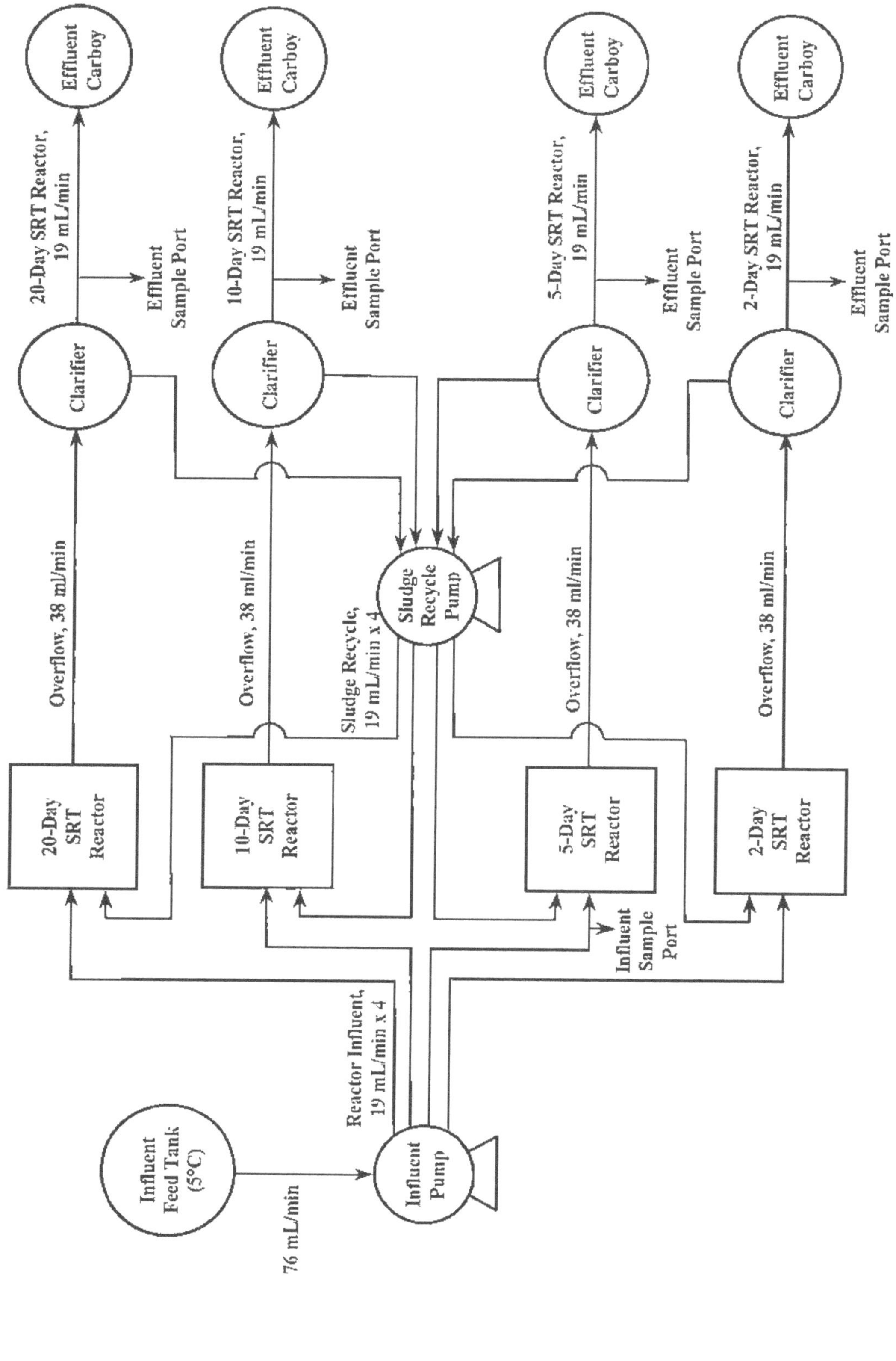

Figure 3-3. Schematic Diagram of Experimental Treatment System.

reactor at the same location and in the same manner as the influent lines. The recycle flow was regulated at 19 mL/min using a second peristaltic pump (Cole Parmer, Inc.) placed below the reactors and operated in the same fashion as the influent pump. Thus configured, each CSTR had an HRT of 8.8 hours, and each clarifier had an HRT of 2.5 hours.

Oxygen was provided to each reactor via two 6" (15.24 cm) long non-clogging porous aquarium aerator tubes using laboratory-provided compressed air. The aerator tubes were mounted to the bottom plate of each reactor via suction cups and placed opposite each other against the sidewalls of the reactors. Prior to reactor introduction, the compressed air was passed through a pressure regulator, filter assembly, and humidifier. The humidifier was constructed from a sealed 20-L carboy half-filled with water. The air entered the humidifier through a submerged tube and bubbled through the water, exiting the humidifier at the top. The air was humidified to prevent excessive evaporation of the mixed liquor in the reactors. The humidified air then passed through a four-way airflow control and metering manifold before flowing to the aerator tubes. In this manner, each reactor's airflow could be separately controlled. This was an important feature when the oxygen requirements for each reactor differed.

3.1.3 DO Control System

In order to monitor and regulate the DO concentration in each reactor, a control system (Great Lakes Instruments) was installed. The system consisted of four DO probes (model 5500-Series), two DO analyzers (model D53), four solenoid valves and four air-blast units. The analyzers and air-blast units were mounted on a 5' x 3', 3/8"-thick plywood base which was placed against the wall of the constant temperature room near the reactors. The in-house compressed air supply was regulated to approximately 15 psi using a constant pressure valve located on the air humidifier.

The four DO probes each contained a hydrophobic membrane, electrolyte, gold anode, silver cathode, and silver reference electrode. The DO in each reactor passed across the hydrophobic membrane and was reduced at the silver cathode using a constant voltage, producing a current between the cathode and anode. This current was directly proportional to the concentration of DO in solution. Each analyzer accepted input from two DO probes that were air calibrated several times throughout the study. To calibrate, each probe was removed from the reactor and rinsed with deionized water then gently wiped to remove any attached growth. The probe was then placed in a sealed plastic bag and allowed to equilibrate to the conditions inside the bag. The analyzer calibrated the probe based upon the DO concentration in the surrounding air.

The air-blast units were initially placed on each of the four probes to regularly air-clean the membranes and minimize biofouling. However, the short blast of air caused a large increase in the DO measurements for several minutes following the blast, and occasionally left air bubbles inside the air-blast unit that disrupted the readings for several hours. This was especially problematic at the lower DO concentration portions of the study. For this reason, the air-blast units were removed and the probes cleaned by hand on a daily basis.

To ensure that the DO control system maintained the targeted level over time via input set points, electronic logging of DO concentrations every 30 seconds was performed. The Highway Addressable Remote Transducer (HART) protocol was used to accomplish this. The HART protocol transmitted output signals from the DO analyzers to a personal computer (PC) via a Bell 202 loop. The Bell 202 loop consisted of the two analyzers, a current-sensing resistor connected in series to a power supply, and an RS232-Bell 202 mini-modem connected to the PC. The software package H-View (Arcom Control Systems) was installed on the PC to read the HART data as it was transmitted. This enabled the PC user to read the DO concentration in each reactor at any given time. In order to save this data throughout the study, a Dynamic Data Exchange (DDE) interface was installed. The DDE interface allowed the data to be viewed in a Microsoft Excel spreadsheet, from which the data was copied and stored in a database using a macro program written in Visual Basic by members of the University of Tennessee's Computer Services Department.

3.1.4 Operating Procedures

To assess treatment performance and bacterial dynamics in this wastewater treatment system, each of the four reactors was maintained at a unique SRT. It is widely accepted that an SRT >3 to 4 days should be employed to achieve complete nitrification in activated sludge for municipal wastewater treatment, and that if nitrification is to be prevented, the SRT should be smaller (Grady et al., 1999). It is also recognized that the SRT required for complete nitrification increases with decreasing temperature and DO concentration (Wild et al., 1971; McClintock et al., 1993; Ydestbo et al., 2000). Therefore, an experimental matrix that included these three parameters was developed and treatment performance was monitored. Bench-scale reactors were operated at SRTs of 20, 10, 5, and 2 days; temperatures of 20°C, 15°C, and 10°C; and DO levels of 2.0mg/L to 3.0 mg/L, 1.0 mg/L to 1.5 mg/L, and 0.2 mg/L to 0.5 mg/L. The 20- and 10-day SRTs were chosen because these values represent the range typically utilized in the local municipal WWTP. The 2-day SRT was chosen because the inverse of this value is close to the typical maximum specific growth rate for nitrifying bacteria. That is, this value represents the point where the nitrifying population would be growing in a zero-order manner, independent of the influent ammonia concentration. The 5-day SRT was chosen to represent a treatment level between highly stressed (2-day) and typical (10-day) operation.

3.1.4.1 Sampling Procedures

In order to maintain these SRTs, solids were wasted directly from each reactor on a daily basis. The wastage volume for each reactor was calculated using the following equation:

$$\text{WastageVolume} = \left(\frac{\text{MLSS} * \text{Volume}}{\text{SRT}} - \text{Q} * \text{ESS} \right) / \text{MLSS} \qquad (3.1.4.1\text{-}1)$$

where:

MLSS = mixed liquor suspended solids (mg/L)
Volume = reactor volume (L) = 10 L
SRT = solids retention time of individual reactor (days)
ESS = effluent suspended solids (mg/L)
Q = influent flow rate (L/day) = 27.36 L/day

This calculation required that solids samples be collected daily to determine MLSS, MLVSS, and ESS concentrations. Samples were obtained in the morning and the values obtained were used to calculate the wastage volume for each reactor on that day. Additional mixed liquor samples were collected from each reactor and stored at -80°C for later molecular analysis. Daily influent and effluent samples were also taken to determine the following process performance parameters: ammonia-N concentration, nitrite-N concentration, nitrate-N concentration, pH, alkalinity, and COD. The daily sampling and wasting procedures were as follows:

1. Effluent grab samples were collected by removing the effluent line from the clarifier ports. Two-hundred fifty milliliters of effluent wastewater from each reactor was then allowed to overflow into two 150-mL Erlenmeyer flasks (125-mL each) placed directly beneath the ports. The two flasks were used to determine nitrogen concentrations, COD, pH, and alkalinity. While the flasks were filling, the temperature in each reactor was recorded.
2. An influent grab sample was then taken by unplugging the influent sampling port plumbed into the 5-day SRT reactor influent line and allowing 250 mL of influent wastewater to flow into two 150-mL Erlenmeyer flasks. Influent samples were taken prior to collection of solids samples so as to ensure that the process had been undisturbed for the maximum amount of time before sampling. The collected influent samples were allowed to acclimate to room temperature before analysis.
3. To ensure that all of the solids for each SRT system were contained in the reactor prior to solids sampling, the influent and recycle pumps, air supply lines, and stir plates were turned off. The reactor effluent lines were then plugged and the recycle pumps turned back on at full speed for approximately 30 seconds, or until all of the solids from the sludge reservoirs and underflow lines were returned to the reactors.
4. Effluent suspended solids (ESS) samples were taken by gently stirring the effluent in the 40-L carboys while pumping the mixed wastewater into 500-mL Erlenmeyer flasks. Four hundred milliliter samples of the 24-hour composite effluent were taken from each of the carboys.
5. The air supply lines and stir plates were then turned back on to ensure that the solids in the reactors were completely mixed prior to sampling. While the reactor solids were mixing, the ESS samples were filtered.
6. Reactor solids were then sampled by removing 20 mL of mixed liquor from the 2-day SRT reactor and 10 mL from the 5-, 10-, and 20-day SRT reactors using a graduated, wide-mouthed serological pipette. These samples were filtered and the filter discs placed in a 105°C oven for 1 hour.
7. While the filters were drying in the oven, sludge volume index (SVI) measurements were taken. Four 1000-mL graduated cylinders were placed in front of the four reactors and filled with mixed liquor using the portable peristaltic pump. The mixed liquor was allowed to settle for 30 minutes.
8. While waiting for the SVI samples to settle, each reactor was cleaned by removing the top plate and scraping any suspended growth from the sidewalls and DO probes with individual wire-bristle brushes and rinsing with deionized (DI) water. The effluent lines were also cleaned as brushes were passed through them and they were rinsed with DI water. Filters were also pre-weighed for the next day's usage.
9. After 30 minutes, the SVI measurements were recorded, and the mixed liquor samples were poured back into their respective reactors. The influent and recycle pumps were then turned back on at full speed for approximately 30 seconds to dislodge any solids that may have settled in the lines. The pumps were then returned to 19 mL/min flow rates.

10. Upon completion of solids sample analysis, the values obtained were used to calculate the wastage volume for each reactor. The volume of mixed liquor already removed from the reactors during sampling was subtracted from the total amount. Except for the 2-day SRT reactor, the total volume of waste sludge was removed in a single wasting event. The waste sludge was pumped directly from the reactors using the portable peristaltic pump and measured using a 2000-mL graduated cylinder. Because of the large volume to be wasted from the 2-day SRT reactor, only half of the total wastage volume was removed at this time. The other half was wasted approximately 12 hours later. After wasting, the volume in each reactor was brought back to 10 L using effluent from that reactor's effluent carboy. Treated effluent was used to restore the reactor volume rather than DI water or tap water to minimize the impact on bacteria in the mixed liquor.

3.1.4.2 Sample Analysis Procedures

As discussed in the previous section, the reactor performance samples were taken just prior to solids sampling to ensure that the reactors had operated for at least 12 hours without disturbance prior to sampling. However, the analysis of the samples occurred simultaneously by two separate student-technicians. Descriptions of the analysis procedures will be given in order of the sampling.

3.1.4.2.1 Ammonium Sample Analysis

Analysis of influent and effluent samples for ammonium concentration was performed in accordance with Standard Method 4500 D, Ammonium-Selective Electrode Method, using an Orion Model 95-12 ammonium-selective probe as suggested by the method (APHA, 1998). One of the two 150-mL Erlenmeyer flasks used to collect grab samples from the clarifier overflow was used from each reactor to analyze for ammonium-N, nitrite-N, nitrate-N, and COD in the system effluent. Also, one of the two 150-mL flasks used to collect grab samples from the influent line was used for the same purpose. One hundred milliliters of effluent were taken from each flask and placed into other 150-mL flasks and allowed to equilibrate to room temperature for ammonium-N analysis. The detection limit for this procedure was 0.1 mg/L as N. The remaining effluent was used for nitrite-N, nitrate-N, and COD analysis, which will be described in the following sections. On days in which it was not possible to analyze samples immediately after collection, samples were preserved by adding 200 μl of concentrated H_2SO_4, covering, and placing in the 5°C cold room where the influent wastewater was stored. Samples were refrigerated no longer than 48 hours prior to analysis.

While the effluent samples were coming to room temperature, 100-mL standards were prepared to calibrate the ammonia-selective probe. The samples and standards had to be at the same temperature to ensure accurate measurements. Standard concentrations of 0.1 mg/L, 0.5 mg/L, 5.0 mg/L, and 50 mg/L ammonium-N were used. Two milliliters of Orion pH Adjusting Solution were added to each standard. The flasks were placed on a small magnetic stir plate, and the probe was placed into the fluid. When the meter readings became stable, they were recorded. The same method was employed for influent and effluent flasks. When all readings were recorded, a standard curve was plotted and ammonium concentrations were read and recorded.

3.1.4.2.2 Nitrite-N and Nitrate-N Sample Analysis

Analysis of influent and effluent samples for the anions nitrite-N and nitrate-N were performed in accordance with Standard Method 4110 B, Ion Chromatography with Chemical

Suppression of Eluent Conductivity using a Dionex DX 500 Ion Chromatograph (IC) with an Iopac AS9HC 4-mm anion exchange column (APHA, 1998). For this procedure, 0.5 mL of the 50 mL of influent and effluent remaining in each flask after the ammonium-N analyses were filtered into Dionex 0.5-mL auto-sample vials using syringes fitted with 0.45-µm glass-fiber filters (Gelman). Samples were analyzed every 3 days (maximum unpreserved sample storage time was 48 hours) to minimize the costs of analysis. The samples not immediately analyzed were stored at approximately 4°C. For each IC run, four standards and one quality-control sample were prepared. Each contained varying concentrations of both anions analyzed in order to calibrate the IC and to ensure accurate measurements. The standards and sample vials were placed into the IC's auto sampler, and the analysis sequence was initiated. The IC automatically plotted a calibration curve (based on the standards) and reported concentrations for each anion in milligrams per liter. These values were then recorded.

3.1.4.2.3 COD Sample Analysis

Analysis of influent and effluent samples for total COD (COD_T) and Soluble COD (COD_S) were performed in accordance with Standard Method 5220 D, Closed Reflux Colorimetric Method (APHA, 1998) using high- and low-range micro-COD test vials (Hach, Inc.). Effluent samples were analyzed for COD_S to eliminate the variability caused by daily fluctuations in the ESS. Approximately 10 mL of the remaining influent wastewater from the ammonium-N samples were placed into 15-mL centrifuge tubes using a syringe without the 0.45-µL filter. Approximately 10 mL of the remaining effluent samples were also placed in 15mL centrifuge tubes using syringes with the 0.45-µL filters. All samples were then preserved with concentrated H_2SO_4 at the recommended 2-mL per liter of sample and stored at 4°C. COD analyses were performed every 2 weeks, well under the 28-day recommended maximum storage time.

The influent and effluent COD tests were performed separately because the influent test required the use of high-range vials (0 to 1500 mg/L), and the effluent test required the low-range vials (0 to 150 mg/L). Five standards were prepared for each test using dried potassium hydrogen phthalate (KHP) to create a standard curve. For each test, 2 mL of sample were placed into a Hach vial, shaken, and digested on a heating block for 2 hours. After cooling, the digested sample was transferred to a clear cuvette and spectrophotometrically analyzed. A standard curve was then plotted and values of COD (mg/L) were recorded.

3.1.4.2.4 Alkalinity and pH Sample Analysis

Analysis of influent and effluent samples for pH was performed using an Orion model 91-57 pH probe. The second 150-mL flasks that contained influent and effluent grab samples from each reactor were allowed to equilibrate with room temperature, and then 50 mL was transferred to another 150-mL flask using a serological pipette. The pH meter was then calibrated using standard solutions of 4, 7, and 10 pH units. Each flask was placed on a small stir plate with the pH probe inserted into the fluid. When the reading became stable, the pH measurement was recorded.

Analysis of influent and effluent samples for alkalinity was performed in accordance with Standard Method 2320 B, Titration Method using the Orion model 91-57 pH probe, and a solution of 0.02 N HCl (APHA, 1998). After the initial pH measurement was recorded, each sample was titrated with 0.02 N HCl to the total alkalinity endpoint (pH=4.5) using an Eppendorf auto-pipette. The volume of acid required to attain this pH was then normalized to $CaCO_3$ and this value was recorded.

3.1.4.2.5 Mixed Liquor Suspended Solids and Effluent Suspended Solids Sample Analysis

Analyses of reactor and effluent samples for MLSS and ESS were performed in accordance with Standard Method 2450 D, total suspended solids (TSS) dried at 103°C to 105°C using 1.5 µm x 47 mm pre-weighed glass fiber filters and a 105°C oven (APHA, 1998). All filters were pre-dried and weighed the day before usage, and pre-wetted with DI water just prior to testing. The 400-mL, 24-hour composite effluent samples were gently poured through the filters and washed with DI water. The filters were then placed in aluminum dishes, dried at 105°C for 1 hour and weighed. The reactor mixed liquor grab samples were filtered, washed with DI water, placed in aluminum dishes, dried at 105°C for 1 hour, and weighed. Values were recorded and used in the following equation to determine MLSS and ESS concentrations:

$$SS = \frac{(W_F - W_I)}{V} \times 1000 \qquad (3.1.4.2.5\text{-}1)$$

where:

SS = suspended solids concentration (mg/L)
W_F = final filter weight (mg)
W_I = initial filter weight (mg)
V = sample volume (mL)

3.1.4.2.6 Mixed Liquor Volatile Suspended Solids Sample Analysis

Analyses of reactor samples for MLVSS were performed in accordance with Standard Method 2540 E, Fixed and Volatile Solids Ignited at 550°C using the dried MLSS filters and a 550°C muffle furnace (APHA, 1998). After the MLSS values were recorded, filters were placed in aluminum dishes and ignited in the muffle furnace for no less than 15 minutes to ignite the volatile portion of the sample. Filters were then removed from the oven and reweighed. Values were recorded and used in the following equation to calculate the MLVSS concentration:

$$MLVSS = \frac{(W_F - W_{FV})}{V} \times 1000 \qquad (3.1.4.2.6\text{-}1)$$

where:

MLVSS = mixed liquor volatile suspended solids (mg/L)
W_F = final filter weight from MLSS analysis (mg)
W_{FV} = final filter weight after MLVSS analysis (mg)
V= sample volume (mL)

3.1.4.2.7 Sludge Volume Index Sample Analysis

Samples used to determine the SVI for each reactor were taken directly from the mixed liquor of each reactor. A modified Standard Method 2710 D, Sludge Volume Index was used to perform this test (APHA, 1998). The method requires a settling column with a stirring mechanism, which was not available for this study. Therefore, a 1000 mL graduated cylinder was used. Activated sludge was allowed to settle in the cylinder for 30 minutes and the settled volume was recorded. These values were then used in the following equation to determine the SVI:

$$SVI = \frac{SSV}{MLSS} \times 1000 \qquad (3.1.4.2.7\text{-}1)$$

where:

SVI = sludge volume index (mL/g)
SSV = settled sludge volume (mL/L)
MLSS = mixed liquor suspended solids (mg/L)

3.2 Industrial Bench-scale Reactor System

3.2.1 Collection of Influent Wastewater

Influent wastewater for the industrial bench-scale system was taken from the neutralization tank at the full-scale WWTP of the Eastman Chemical Company (Kingsport, TN). Wastewater was transported from the sample collection point to the laboratory using two 55-gallon (208.2 L) steel drums then pumped into a 110-gallon influent feed tank. Each batch of industrial wastewater was sufficient to feed the bench-scale treatment system for 3 to 4 days.

3.2.2 Experimental System and Operation

The industrial bench-scale wastewater treatment system was designed to simulate Eastman's full-scale WWTP. The bench-scale system contained four separate tanks, each of which consisted of three 10-L baffled reactors operating in series (Figures 3-4 and 3-5). Three pumps were used to supply influent wastewater to each of the three stages of the four reactor trains (tanks) in the same proportion as that found in the full-scale WWTP. The first-, second-, and third-stage reactors in each treatment train receive approximately 50%, 30%, and 20% of the influent flow, respectively: 10.7 L/d to the first-stage reactors, 6.4 L/d to the second-stage reactors, and 4.3 L/d to the third-stage reactors. The total influent flow rate was 21.4 L/d/train, or 85.6 L/day for the entire treatment system. This flow rate yields a hydraulic retention time of 33.6 hours per train; 22.4 hours in the first stage, 14 hours in the second stage, and 11.2 hours in the third stage reactors. Effluent from each treatment train overflowed to a 1.9-L external secondary clarifier that provided quiescent settling and continuous recycle of the settled sludge. A recycle pump transferred settled sludge back to the first stage of each train. Each reactor was equipped with a micro-submersible pump that provided complete mixing within the tank. Aeration for all reactors was supplied as compressed air though diffuser tubing in the bottom of each reactor. A large stand was used to support the reactors and all ancillary equipment. This allowed the entire system to be placed on a bench top in a laboratory located at the WWTP site.

A DO monitor and control arrangement, similar to the one used in the municipal system, was installed in the industrial bench-scale system to maintain the DO concentration at a constant level. The system consisted of 12 DO probes (Model 5500-Series), six DO analyzers (Model D53), and 12 air-blast units. The DO level in the industrial system was controlled similarly to the municipal system.

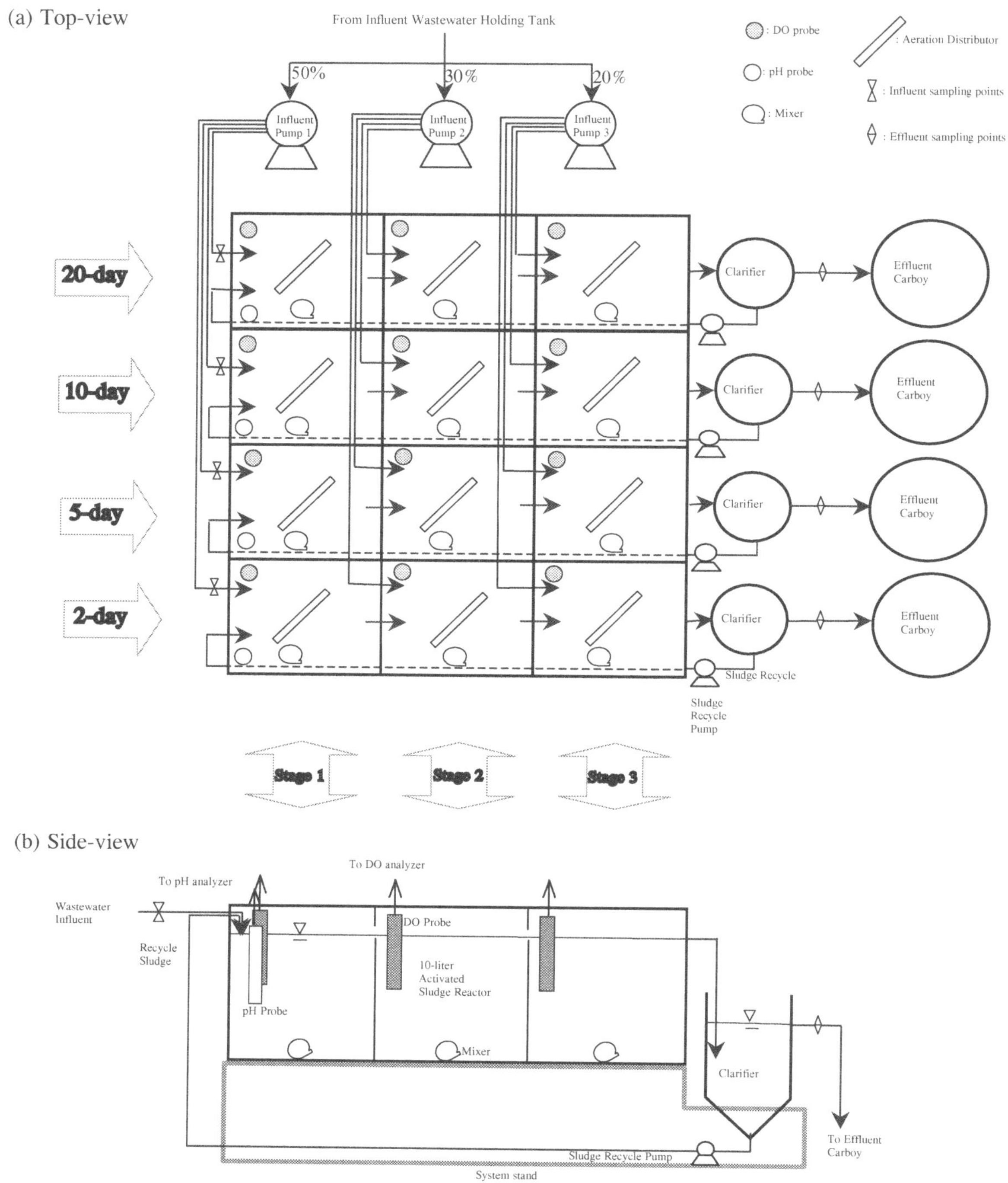

Figure 3-4. Top-view (a) and Side-view (b) Diagrams of the Industrial Bench-scale System.

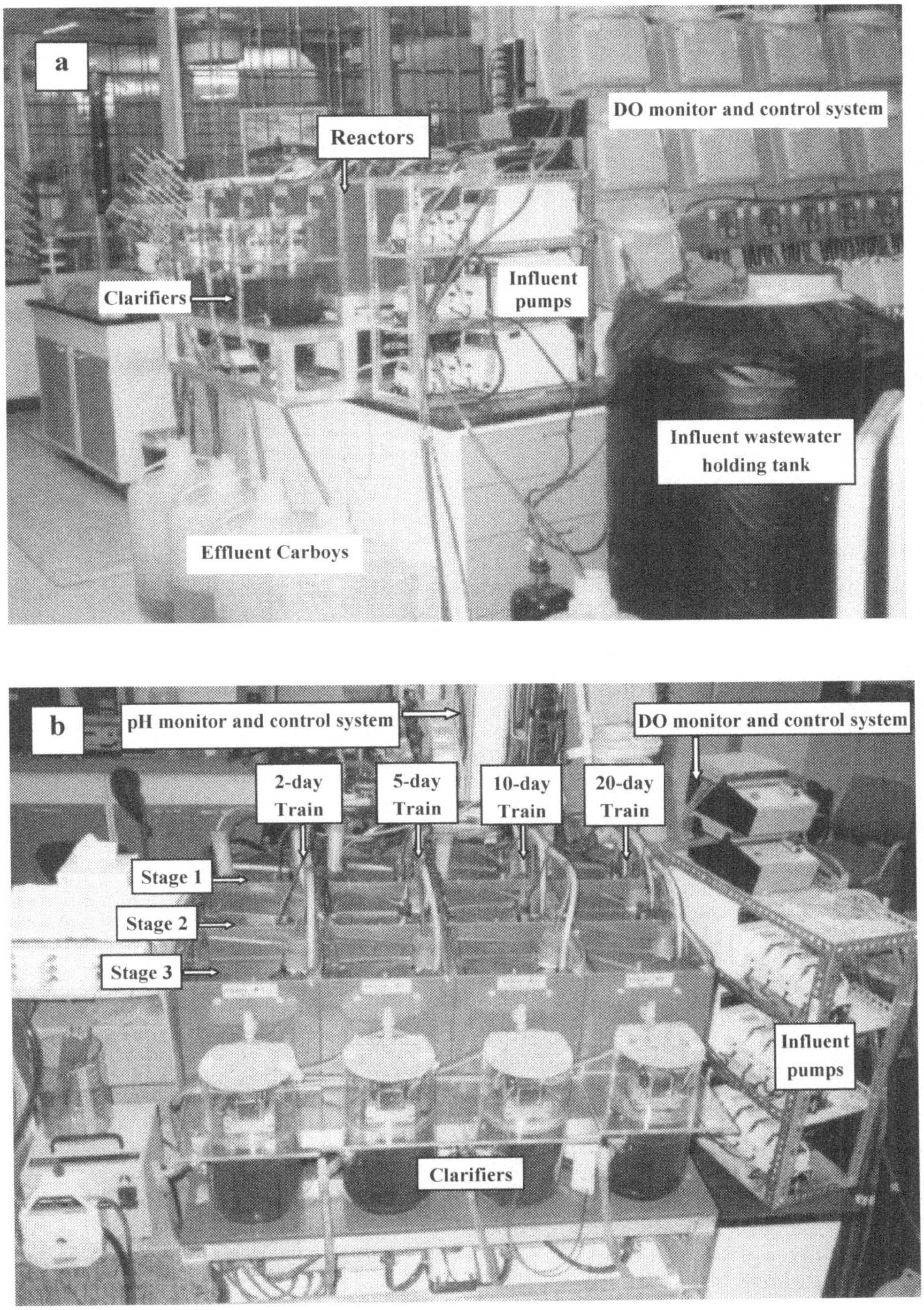

Figure 3-5. Industrial Bench-scale Activated Sludge System Consisting of Influent Wastewater Holding Tank, Influent Pumps, Reactors (4 trains × 3 stages), Clarifiers, Effluent Carboys, pH Monitor and Control System, and DO Monitor and Control System.

Additionally, a pH monitor and control system was installed prior to Phase 2 operation (ammonia addition) in the industrial bench-scale system to maintain pH. The pH system consisted of four Walchem WDP300 Series pH/ORP controllers, four pH probes, and four pumps. A solution of sodium hydroxide (NaOH) was used to maintain pH within a neutral range (7.0 to 7.2). The control system was designed to add the NaOH solution to the first stage reactors of each train when the pH dropped below 7.0 and to stop the NaOH flow (turn off) when the pH reached 7.2.

The industrial bench-scale system was operated at room temperature except during the 2-week high temperature trial. During this period, heating coils were installed in the first and second stage of each train. Because the heating coils could not maintain the temperature in the reactors at a constant level, the high temperature trial was stopped and all eight heating coils were removed. Each of the four trains in the industrial system was operated at a unique SRT (20, 10, 5, and 2 day). The overall study period consisted of four operating phases in which two DO levels (high and low) and two ammonia loading levels (high and low) were tested.

3.2.3 Sampling and Analysis

The sampling and analysis procedures for the industrial bench-scale system were similar to those outlined for the municipal bench-scale treatment system. Analysis items for each industrial sample are summarized in Table 3-1. Influent and effluent samples were collected 5 days per week from each SRT train of the industrial bench-scale system to assess treatment performance. The influent sampling point was located upstream of the first stage for all four SRT trains. The effluent sampling point was located downstream of each clarifier.

Table 3-1. Analyses Performed on Samples Taken from the Industrial Bench-scale System.

Samples	Analyses										
	TKN	NH_4^+	NO_2^-	NO_3^-	COD	Alkalinity	MLSS	MLVSS	Eubacteria	*N. nitrosa*	*Nitrospira*
Influent	√	√	√	√	√	√					
Solids							√	√	√	√	√
Effluent	√	√	√	√	√	√	√	√			

One additional analysis (TKN) was performed by personnel at Eastman Chemical Company on the industrial wastewater. A total of 10 TKN samples (one influent and four effluent samples per day twice a week) were collected, filtered, acidified, and stored at 4°C before analysis. Nitrite and nitrate samples were collected, filtered, and stored at 4°C for analysis twice per week. Ammonia and alkalinity samples were analyzed every sampling day. The COD samples were filtered, acidified, and stored at 4°C for analysis every 2 weeks. The concentrations of MLSS and MLVSS were measured daily in each of the 12 reactors to determine the wastage volume for each SRT train. Additionally, solids concentrations were determined for effluent samples taken from the collection container for each SRT system. Daily biosolid wastage volumes were determined based on solids concentrations in the mixed liquor and effluent of each

SRT treatment train. Biosolid samples were also collected from each reactor and stored at -80°C for subsequent molecular analysis.

The analysis methods for ammonia, nitrite, nitrate, COD, alkalinity, MLSS, and MLVSS were the same as described for the municipal samples. Analysis of influent and effluent samples for TKN was performed in accordance with U.S. Environmental Protection Agency (EPA) Method 351.2: Automated Determination of Total Kjeldahl Nitrogen in Water and Wastewater.

3.3 Molecular Methods

3.3.1 Evaluation and Selection of Molecular Techniques

The overall goal of this research was to quantify microbial nitrifying activity under varying operational conditions during activated sludge treatment using molecular-based techniques. The advantage of molecular methods over traditional methods is that microorganisms can be identified and monitored without cultivation. During the first 2 years of this project, various molecular methods were evaluated for the detection of nitrifying bacteria in activated sludge (summarized in Table 3-2). The nucleic acid hybridization methods (RNA with 16S rRNA oligonucleotide probes and DNA with catabolic gene probes) were abandoned because the probe targets in the activated sludge were below the detection limits. PCR techniques were then assessed. The application of PCR methodologies to measure nitrifying bacteria in the activated sludge was found to have two advantages over hybridization methodologies. First, the detection limits for PCR methodologies are one to two orders of magnitude lower than hybridization methodologies. Second, the PCR product can be cloned and sequenced to verify that the measured target is correct, whereas in hybridization experiments the measured target cannot be verified.

Table 3-2. Molecular Methods Evaluated for the Quantification of Nitrifying Bacteria in Activated Sludge.

Method Evaluated	Conclusion
Fluorescent In Situ Hybridization (FISH)	FISH provides good qualitative data, but is difficult to use for high throughput quantification of nitrifying bacteria in activated sludge.
Solution Hybridization	Has the same problem with sensitivity as with filter hybridizations.
DNA hybridization with *amo* and *nor* gene probes (Catabolic probes)	*amo* and *nor* genes were below the detection limit of 0.01 ng/µg DNA (<5% of total population).
RNA hybridization with oligonucleotide 16S rDNA probes (16S probes)	Ammonia- and nitrite-oxidizing bacteria specific 16S rDNA sequences were below the detection limit (≤1% of total 16S rDNA).
Competitive PCR	Effective (data published in Dionisi et al., 2002a,b); however, time-consuming, expensive, and inflexible.
Real-time PCR with Molecular Beacon probes	Effective (data published in Harms et al., 2003); however, more costly and more difficult to design than Taq probes.
Real-Time PCR with TaqMan probes	Effective, chosen method for analysis of full-scale plant (Harms et al., 2003) and bench-scale reactor systems.

3.3.1.1 Nucleic Acid Hybridizations

Ribosomal RNA and messenger RNA (mRNA), along with DNA, are the primary targets for nucleic acid hybridization. This method is based on the ability of two complementary single-stranded DNA molecules or single-stranded DNA and RNA molecules to form a hydrogen-bonded duplex structure. Fundamentally, target nucleic acid (unknown DNA or RNA) extracted from environmental samples is bound in the single-stranded form (after denaturation in the case of DNA) to a solid support such as a nylon membrane. At appropriate pH and temperature conditions, a labeled nucleic acid probe (DNA or RNA, also in single-stranded form) is added to the reaction mixture. Under the proper conditions, the single-stranded probe forms base pair hydrogen bonds only with those sequences that are homologous to the complementary base sequences of the target nucleic acid (Atlas et. al., 1992).

3.3.1.1.1 DNA Hybridizations with Catabolic DNA Probes

This method consists of applying appropriate amounts of total DNA extracted from activated sludge to a nylon membrane, together with a set of DNA standards that allows the generation of a standard curve to quantify the target DNA. The DNA is hybridized on the membrane with an appropriate gene probe, followed by detection and quantification (Applegate et al., 1995).

Genes encoding specific enzymes of AOB and NOB were amplified from pure cultures using PCR. Primers specific for the amplification of hydroxylamine oxidoreductase (*hao*) and ammonia monooxygenase (*amo*) genes from *N. europaea*, and *nor*B from *Nitrobacter hamburgensis* were designed using the program Primer 3 (www.genome.wi.mit.edu) (Table 3-3).

Table 3-3. Primers for the Amplification of *hao*, *amo*A and *amo*AB from *N. europaea*, and *nor*B from *N. hamburgensis*.

Primer	Sequence
hao-f	5' GAGTGGATGAGGGGTTTATTGC 3'
hao-r	5' TTGCCTTCCAGTTTGTTTACACG 3'
amoA-f	5' AGAAATCCTGAAAGCGGC 3'
amoA-r	5' GATACGAACGCAGAGAAG 3'
amoAB-f	5' TGGATGGACTGGAAAGATCG 3'
amoAB-r	5' TCGAAATTCAACCGACAGC 3'
NO166-f	5' CAAGTTTCGATGGTCTTTCACC 3'
NO1419-r	5' CTTCTTCGAGCGCATATAGACG 3'
NO325-f	5' GACTATCAGAACCTGATCAACG 3'
NO258-r	5' GCTGAGCAATATCTTCTACAACC 3'
NO1302-r	5' AATGTCCTTGACCTTCTTCG 3'

The PCR products corresponding to *amo*A, *amo*AB and *hao* genes from *N. europaea*, and the gene *nor*B from *N. hamburgensis* were cloned into a plasmid vector in *Escherichia coli* using a TOPO™ TA cloning kit (Invitrogen, Carlsbad, CA). The plasmid DNA for each clone was purified and the amplified product was sequenced in order to confirm the identity of those genes.

These plasmids were used to make stock solutions at different concentrations for preparing a standard curve for each gene.

DNA from activated sludge samples (full-scale municipal, full-scale industrial, or bench-scale reactors) was extracted using the FastDNA kit (Bio101, Vista, CA). An internal standard DNA was added to the sample (λ DNA) to measure the extraction efficiency. DNA from each sample and standard were immobilized on nylon membranes for slot blot hybridization experiments. The membranes were hybridized with the specific radiolabeled probe, and the signal obtained in each sample was compared with that obtained in the standards. Slot blot using λ DNA as a probe was utilized to measure the DNA extraction efficiency. The FastDNA kit (Bio101, Vista, CA) showed high extraction efficiency, between 85% and 100%.

The appropriate hybridization and washing conditions were adjusted for each probe. The probes *amo*A and *amo*AB from *N. europaea* hybridized poorly with the activated sludge samples, suggesting that AOB were present in the activated sludge in a proportion below the detection limit of the technique, or that the *amo* genes of the ammonia-oxidizers in the activated sludge are not highly conserved with respect to the *amo* genes from *N. europaea*. In order to evaluate these possibilities, the gene *amo*A was cloned from the activated sludge (see below) and one clone was selected for use as a probe. A hybridization analysis of seven samples collected on different days from the 2-day SRT reactor (loaded in triplicate and in three DNA concentrations) indicated that the amount of *amo*A in the sample was below the detection limit of this technique. The probe *nor* from *N. hamburgensis* was also evaluated with DNA samples from the reactors. The amount of *nor* in the sample was also below the detection limit.

3.3.1.1.2 RNA Hybridization with 16S rDNA Oligonucleotide Probes

The quantitative dot blot hybridization method using 16S rRNA targeted oligonucleotide probes can be used to determine the relative rRNA abundance of a specific bacterial group relative to total rRNA, as long as it is above the detection limit. RNA extracted from wastewater biosolids and pure cultures of ammonia- and nitrite-oxidizing bacteria immobilized on nylon membranes, was hybridized with P^{32} labeled universal and specific oligonucleotide probes. The oligonucleotide probes and target bacterial groups tested with activated sludge samples are listed in Table 3-4.

Two methods for RNA extraction from the activated sludge were tested: a hot phenol method designed in our laboratory (Fleming et. al. 1992) and the FastRNA kit (BIO101, Vista, CA). Dot blot hybridization and gel electrophoresis of RNA extracted from activated sludge samples using both methods indicated that the RNA yield for the extraction kit method was insufficient for use in hybridization experiments. Therefore, RNA was extracted using the hot phenol method from activated sludge originating from the bench-scale reactors and the full-scale municipal WWTP, and from pure cultures of the control strains listed in Table 3-5.

The hybridization temperatures for each oligonucleotide probe were optimized using standard RNA from pure cultures with zero, one, or two mismatches to the oligonucleotide probe. In addition, the detection limit for each probe was determined by adding different amounts of standard RNA to the nylon membrane. One base pair mismatch could not be clearly discriminated from the zero mismatch control for any of the oligonucleotide probes tested, because as the hybridization temperature increased, the signals obtained for the positive control RNA became fainter. The optimal hybridization and wash temperatures for each probe are listed in Table 3-4.

Table 3-4. Oligonucleotide Probes and Their Target Organisms.

Oligonucleotide probe	Sequence of probe (5'-3')	Target organism	Optimal Temp. (°C)	Reference
UNIV1392	ACGGGCGGTGTGTRC	Universal	43	Olsen et al. (1986)
NEU	CCCCTCTGCTGCACTCTA	Most halophilic and halotolerant *Nitrosomonas*	64	Wagner et al. (1995)
NSO1225	CGCCATTGTATTACGTGTGA	All ammonia-oxidizing bacteria within the β-subgroup of *Proteobacteria*	50 - 55	Mobarry et al. (1996 and 1997)
NSO190	CGATCCCCTGCTTTTCTCC	All ammonia-oxidizing bacteria within the β-subgroup of *Proteobacteria*	53 - 55	Mobarry et al. (1996)
NIT3	CCTGTGCTCCATGCTCCG	*Nitrobacter* sp.	50 - 55	Wagner et al. (1996)
S-*-Ntspa-1026-a-A-18	AGCACGCTGGTATTGCTA	*Nitrospira moscoviensis*	Not tested	Juretschko et al. (1998)

Table 3-5. Pure Cultures of Bacteria Used as Controls for the Optimization of Dot Blot Hybridization Conditions.

Microorganism	Function
Nitrosomonas europaea	positive control for probes NEU, NSO1225, and NSO190
Nitrobacter winogradskyi	positive control for probe NIT3
Nitrobacter hamburgensis	positive control for probe NIT3
Comamonas testosteroni	1 mismatch control for probe NEU
mXyN1 (denitrifier, β-subgroup of proteobacteria; Rabus et al. 1995)	2 mismatch control for probe NEU
Rhodopseudomonas palustris	1 mismatch control for probe NIT3
Bradyrhizobium japonicum	2 mismatch control for probe NIT3
Spirillum volutans	2 mismatch control for probe NSO1225; 1 mismatch control for probe NSO190

Dot blot hybridizations were performed with RNA extracted from the reactor activated sludge and the oligonucleotide probes listed in Table 3-4, except the S-*-Ntspa-1026-a-A-18 probe. This probe was not tested because the control strain (*Nitrospira moscoviensis*) was not available. Hybridization with probe UNIV1392 indicated that sufficient total RNA from each reactor was added to the nylon membrane. In all cases, the hybridization signals of RNA from the bench-scale municipal reactors with the specific probes were weak and less than the detection limit of 10 ng RNA. These results indicated that RNA dot blot hybridization with specific oligonucleotide probes was not sensitive enough for this project. Based on our blots, the percent of total 16S rRNA from ammonia and nitrite-oxidizing bacteria in the activated sludge samples must be ≥1% in order to confidently distinguish between these bacteria and other bacteria having 16S rRNA sequences with one mismatch in the probe region. This detection limit is in the published range of 0.1% to 1% for oligonucleotide probes (Amann et al., 1995).

3.3.1.2 PCR Methods for Quantification of Nitrifying Bacteria

PCR methodologies were evaluated for use in quantifying nitrifying bacteria due to the low sensitivity and lack of specificity associated with dot blot hybridizations. Two methods can be used to quantify PCR products from environmental samples: 1) competitive PCR; and 2) real-time PCR. Both methods were evaluated for quantifying *amo* gene targets and 16S rDNA sequences from nitrite-oxidizing bacteria in activated sludge samples. In order to design PCR assays, sequence information for nitrite oxidizers and ammonia oxidizers were obtained from the full-scale municipal and industrial WWTPs and the bench-scale treatment systems and compared to sequences in the database (e.g., GenBank).

3.3.1.2.1 DNA Extraction

In order to obtain sequence information and perform PCR analysis, genomic DNA was extracted from mixed liquor samples ranging in volume from 1 mL to 25 mL, depending on the MLVSS values, using a FastDNA kit (BIO 101, Vista, CA) with minor modifications. These modifications included washing the binding matrix-DNA complex twice with 80% (vol/vol) ethanol after the recommended salt-ethanol wash step and eluting the DNA in 100 µL of 10 mM Tris-HCl buffer (pH 8.0) (Dionisi et al., 2002). The integrity of the DNA samples was analyzed by electrophoresis in 0.8% (wt/vol) agarose (Fisher Scientific, Pittsburgh, PA), 1X TBE and 1X Gelstar® nucleic acid gel stain (FMC Corporation, Rockland, ME). The DNA was quantified by using a DyNA Quant200 fluorometer (Hoefer Pharmacia Biotech, San Francisco, CA).

3.3.1.2.2 PCR Amplification of Nitrospira 16S rDNA from Activated Sludge

The target molecule for detecting the nitrite-oxidizing bacteria *Nitrospira* was the 16S rDNA. This molecule was chosen because the genes for nitrite oxidation have not been cloned and sequenced from *Nitrospira*. In addition, the *Nitrospira* found in activated sludge appear to be monophylogenetic and conserved between municipal and industrial WWTPs (Burrell et al., 1998; Dionisi et al., 2002). Two primers were designed (NSR1113f and NSR1264r) (Table 3-6) and tested using genomic DNA extracted from municipal and industrial biosolids as templates. PCR amplification of the *Nitrospira* sp. 16S rDNA was performed in 25-µL (total volume) mixtures containing Ready-To-Go PCR beads (Amersham Pharmacia, Piscataway, NJ), 10 pmol of primer NSR1113f, 10 pmol of primer NSR1264r, and 2ng of genomic DNA from MLSS samples as the template. The program for amplification was 5 minutes at 94°C; followed by 40 cycles consisting of 30 seconds at 94°C, 30 seconds at 65°C, and 30 seconds at 72°C; and a final cycle consisting of 15 minutes at 72°C. Plasmid DNA from several randomly selected clones were isolated with an RMP AFS kit (Bio 101) and completely sequenced. All DNA molecules were sequenced at the Molecular Biology Resource Facility of the University of Tennessee (Knoxville) using an ABI PRISM dye terminator cycle sequencing kit with Ampli Taq DNA polymerase (protocol P/N 402078, revision A), an Applied Biosystems 373 DNA sequencer

Table 3-6. Target Molecules and the Associated PCR Primers/Fluorescent Probes Used to Detect Nitrifying Bacteria.

Target molecule	Forward and reverse primers	PCR method (sequence for TaqMan probes)
Nitrospira 16S rDNA	NSR1113-5'-CCTGCTTTCAGTTGCTACCG-3' NSR1264r -5'-GTTTGCAGCGCTTTGTACCG-3'	Competitive PCR, Molecular Beacon, TaqMan NtspaTaq; 5'-6-FAM- AGCACTCTGAAAGGACTGCCCAGG-TAMRA3'
*N. oligotropha amo*A (*N. marina* cluster)	*amo*598f -5' GAATATGTTCGCCTGATTG3' *amo*718r -5'CAAAGTACCACCATACGCAG3'	Competitive PCR
*N. oligotropha amo*A (*N. marina* cluster)	*amo*No550D1f -5'-TCAGTAGCTGACTACACMGG3' *amo*No550D2f-5'- TCAGTAGCYGACTACACMGG3' *amo*No754r-5'-CTTTAACATAGTAGAAAGCGG-3'	*amo*NoTaq716r 5'-6-FAM- CCAAAGTACCACCATACGCAG-TAMRA-3'
*N. nitrosaamo*A (*N. europaea* cluster)	*amo*Nn549f-5'-TTTCAATGGCAGACTACA3' *amo*Nn777r -5'-TTTAACAATACGGCCTCT3'	*amo*NnTaq7405'-6-FAM- AATGCTGTACAGTAAACTTTACCGAG- TAMRA-3'
Eubacteria 16S rDNA	1055f -5'-ATGGCTGTCGTCAGCT3' 1392r -5'-ACGGGCGGTGTGTAC3'	16STaq1114f 5'-6-FAM- CAACGAGCGCAACCC-TAMRA-3'

(Perkin-Elmer, Foster City, CA) and M13f or M13r located on the pCR 2.1 plasmid as the primer. The reactions produced a fragment with the expected size of approximately 150 base pairs (bp). These PCR products were cloned, and a total of 18 clones were randomly selected and sequenced. All sequences were identified as *Nitrospira*, with greater than 98% similarity to each other, suggesting that NOB in activated sludge are less diverse and phylogenetically distinct than AOB (Dionisi et al., 2002). No false positives were detected, indicating a high specificity of these primers for the amplification of 16S rDNA from *Nitrospira*. Because of the similarity of *Nitrospira* 16S rDNA, this assay could be applied to different activated sludge systems without further optimization (Dionisi et al., 2002).

3.3.1.2.3 PCR Amplification of the amoA Gene from Activated Sludge

The *amo*A gene represents a very powerful molecular tool for analyzing ammonia-oxidizing communities due to i) its specificity, ii) its fine-scale resolution of closely related populations, and iii) because it is a functional trait rather than a phylogenetic trait (Rotthauwe et al., 1997). The gene encoding the enzyme ammonia monooxygenase (*amo*A) was amplified using the primers *amo*A1-F and *amo*A2-R (Rotthauwe et al., 1997), and the DNA isolated from the full-scale municipal and full-scale industrial WWTP activated sludge as a template. PCR amplification was performed in 25-µL (total volume) mixtures by using Ready-To-Go PCR beads (Amersham Pharmacia, Piscataway, NJ). The program for amplification was 5 minutes at 94°C; followed by 50 cycles consisting of 30 seconds at 94°C, 30 seconds at 62°C, and 30 seconds at 72°C; and a final cycle consisting of 7 minutes at 72°C. The amplified products were cloned into the pCR2.1 vector using a TOPO™ TA cloning kit (Invitrogen, Carlsbad, CA) according to the instructions of the manufacturer to generate an *amo*A library. Clones were screened by colony hybridization as previously described (Sambrook et al., 1989), using the *amo*A gene from *Nitrosomonas europaea* as the probe (McTavish et al. 1993).

The sequences of the clones were aligned with the DNA sequences for the *amo*A gene from pure cultures using the program Clustal X, to produce a phylogenetic tree (Figure 3-6). Phylogenetic trees are a specific type of relational graphic representation reflecting genealogical

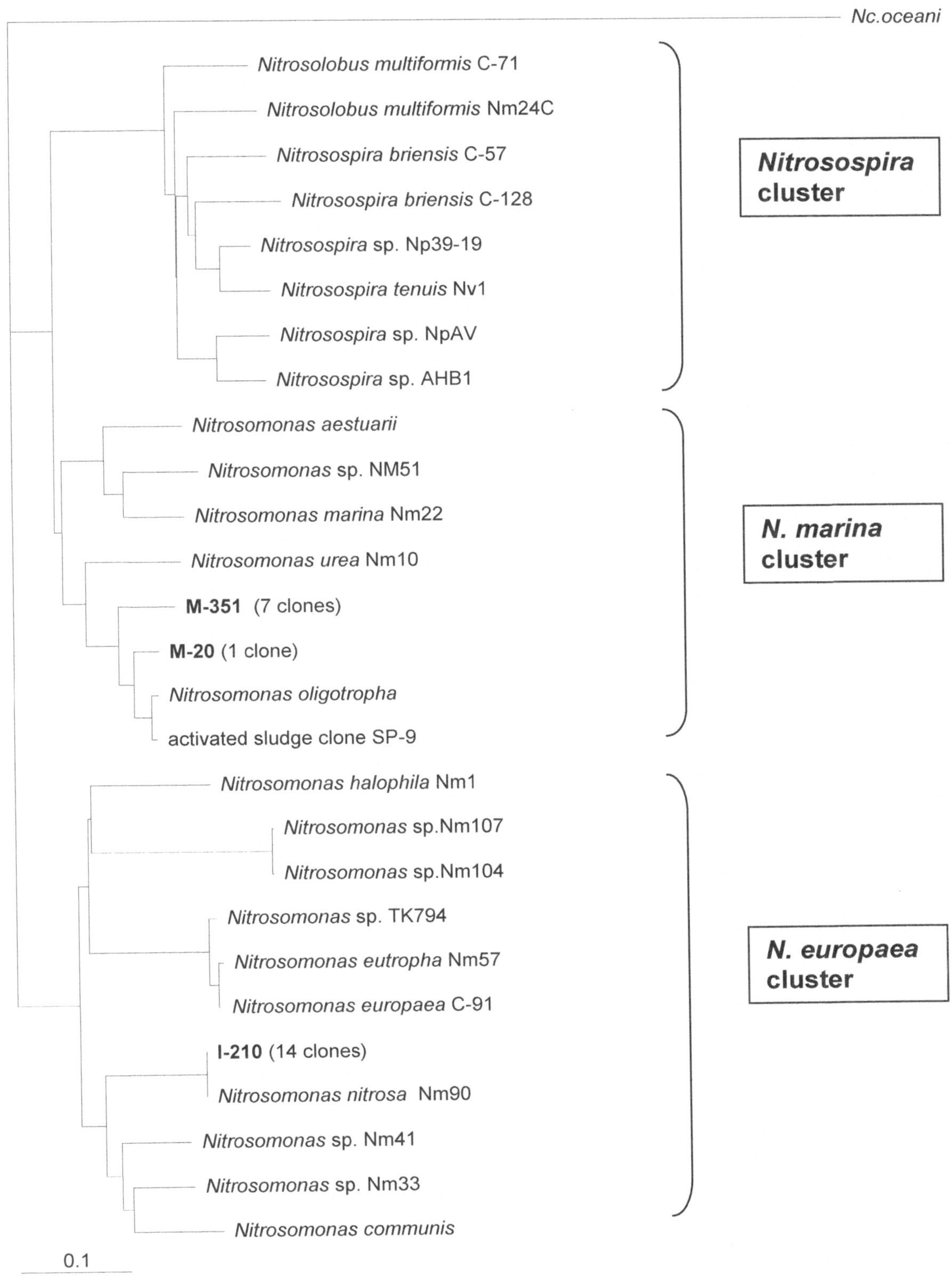

Figure 3-6. Phylogenetic Fitch-Margoliash amoA Dendogram.

or evolutionary connections in a group of organisms (Hillis et al., 1993). The *amo*A phylogenetic tree shows the presence of three clusters: the γ-proteobacteria cluster represented by *Nitrosococcus oceanus*, and two different clusters that belong to β-proteobacteria: *Nitrosomonas* (including *Nitrosomonas* spp. and *Nitrosococcus mobilis*) and *Nitrosospira* (including the genera *Nitrosospira*, *Nitrosovibrio*, and *Nitrosolobus*). The *amo*A clones from the activated sludge are more closely related to the *Nitrosospira* cluster. The alignment of the DNA sequences of the *amo*A clones with the published *amo*A gene sequence from *N. europaea* showed between 80% and 90% similarity. In contrast to the *Nitrospira* assay, designing a PCR assay to detect AOB is more complicated for two reasons. First, WWTPs may contain multiple species of AOB (Purkhold et al., 2000). Second, most AOB are phylogenetically closely related to other activated sludge bacteria within the β-subdivision class of Proteobacteria. Therefore, several primer sets are needed to perform ecological studies of different AOB populations in activated sludge (Dionisi et al., 2002). Due to sequence diversity in the *amo*A gene found in the full-scale municipal and industrial WWTPs and the bench-scale treatment systems, a single *amo*A PCR assay could not be designed. Therefore several primer pairs were needed and were modified over the course of the research project to improve specificity and sensitivity.

3.3.1.2.4 Competitive PCR

Competitive PCR assays were first designed for this project because the cPCR procedure is a well-established quantitative method used to enumerate both culturable and nonculturable organisms, including nitrifying bacteria (Johnsen et al. 1999, Kowalchuk et al. 1999, Mendum et al. 1999, Stephen et al. 1999). Competitive PCR is based on the measurement and comparison of two different templates during the reaction. One template is the original gene fragment (from a DNA sample extracted from the activated sludge). The second template, known as the competitor sequence, is the same gene fragment containing a deletion or insertion and serves as an internal standard. The competitor sequence is co-amplified in competition with the specific DNA sequences in the sample. This DNA fragment can be separated from the target product on an agarose gel, because of its size difference (Johnsen et al. 1999, Mendum et al. 1999, Stephen et al. 1999). Each competitive PCR assay consists of multiple reaction mixtures, including serial dilutions of 10^6 to 10^1 competitor molecules and aliquots of sample DNA. The band intensities are analyzed and quantified with a gel imaging system.

A 119 bp oligonucleotide consisting of a 32 bp deletion in the 151 bp target sequence of the *Nitrospira* spp. The 16S rDNA was constructed and cloned into *E. coli* for use as the competitor (Dionisi *et al.*, 2001). The 16S rDNA amplification was performed using Ready-To-Go PCR Beads, 10 pmol of each primer, 0.1 ng to 0.5 ng of biosolids genomic DNA, and 2500 to 100,000 copies of the competitor DNA. Amplification was performed using a initial step of 5 minutes at 94°C; followed by 35 to 40 cycles consisting of 30 seconds at 94°C, 30 seconds at 65°C, and 30 seconds at 72°C; and a final cycle of 15 minutes at 72°C. PCR products were separated by electrophoresis in 3% (wt/vol) NuSieve 3:1 agarose, 1X TBE and 1X GelStar nucleic acid gel stain. The intensities of the individual bands were quantified using an AlphaImager™ 1220 documentation and analysis system. The fluorescence of the competitor product was corrected by a factor of 151/119. The determination of the equivalence point was performed as previously described (Chandler, 1998). Competitive PCR assumes equal amplification efficiency of target and competitor in order to obtain an accurate measure of the number of targets in the reaction. However, in control experiments the target amplified 2.74 times more efficiently than the competitor sequences. Similar differences in amplification efficiencies (3.1 fold) have been reported for other competitive PCR systems (Lee et al., 1996). An additional correction factor of 1/2.74 was applied for calculating the copy number values for all biosolids samples.

A 100-bp oligonucleotide corresponding to the target DNA (121-bp of the *amo*A gene) with an internal deletion of 21-bp was synthesized and cloned into *E. coli* (Dionisi et al., 2001). The cloned DNA sequence was amplified and quantified for use as the competitor in the competitive PCR sample analysis. For sample analysis, amplification was performed using Ready-To-Go PCR Beads (Amersham Pharmacia, Piscataway, NJ),10 pmol of *amo*598f and *amo*718r primers, dilutions of the competitor (from 2500 to 100,000 copies) and 15 ng to 30 ng of suspended solids genomic DNA. The program used for amplification was 5 minutes at 94°C; then 30 cycles consisting of 15 seconds at 94°C, 15 seconds at 60°C, and 15 seconds at 72°C; and a final cycle of 7 minutes at 72°C. PCR products were separated by electrophoresis in 3% (wt/vol) NuSieve 3:1 agarose (BMA, Rockland, MD), 1X TBE (Sambrook et al., 1989), and 1X GelStar® nucleic acid gel stain (FMC Corporation, Rockland, ME). The intensities of the individual bands were quantified using an AlphaImager™ 1220 documentation and analysis system (Alpha Innotech Corporation, San Leandro, CA). The fluorescence of the competitor product was corrected by a factor of 121/100. The determination of the competition equivalence point was performed as previously described (Chandler, 1998).

The expected copy number and the determined copy number were in agreement for the *amo*A gene. The efficiency of amplification of the competitor and target DNA was the same between 10 and 50 cycles of the PCR reaction. When the *amo*A competitive PCR technique was applied to DNA from a biosolids sample, only two bands, competitor and target, were detected in the gel. The addition of 15 ng to 30 ng of biosolids genomic DNA to the PCR reaction produced reproducible quantitative PCR analysis. However, when more than 50 ng of DNA was added, amplification of both competitor and target DNA was inhibited. Based on our results, the competitive PCR assay was suitable for use in a complex DNA sample, and was capable of detecting and quantifying as low as 2500 copies of the *amo*A gene in the PCR reaction. However, competitive PCR is laborious and expensive because a minimum of 6 PCR reactions is needed per sample analyzed. Therefore, the number of samples that can be analyzed is limited. In addition, a specific competitor sequence similar to the target sequence must be constructed for each target sequence making the system less flexible.

3.3.1.2.5 Real-Time PCR

More recently developed real-time PCR methods were investigated as an alternative to cPCR because they combine high throughput with high analytical sensitivity and precision and are less expensive than cPCR. In real-time PCR, amplicons are detected by measurement of a fluorescence signal without post-PCR sample processing, such as gel electrophoresis. Two types of fluorescent probes commonly used in real-time PCR are Molecular Beacon (Tyagi and Kramer, 1996; Broude, 2002) and TaqMan (Livak et al., 1995).

3.3.1.2.6 Molecular Beacon PCR

The molecular beacon method is fast and precise. It can detect seven copies of a gene within a sample (Piatek et al., 1998; Vet et al., 1999; Kostrikis et al., 1998; Tyagi and Kramer, 1996). Molecular beacon PCR (Cayouette et al., 1999; Kostrikis et al., 1998; Tyagi and Kramer, 1996; Tyagi et al., 1998; Tyagi et al., 1999) allows measurement and quantification of PCR products online during PCR amplification using a thermocycler equipped with a fluorimeter.

PCR product formation is measured by the fluorescence emitted by the fluorophore on the molecular beacon when the beacon binds to the target sequence. The molecular beacon is added with the primers and template DNA at the start of the PCR reaction.

A molecular beacon is a hairpin-shaped hybridization probe, composed of a loop (probe sequence) and a stem (composed of the two complementary arms of the beacon). A fluorophore and non-fluorescent quencher moiety are attached on opposite arms. When the beacon is free in solution, it has a hairpin structure; however, when the fluorophore and quencher are close together, the energy of the fluorophore is quenched. If the complementary target is present, the beacon binds to the target, thereby removing the hairpin structure. This separates the quencher from the fluorophore, and fluorescence is emitted. Therefore, the amount of fluorescence is dependent upon the amount of specific PCR product formed. Fluorescent measurement occurs at the annealing step of each PCR cycle. The initial copy numbers of a specific sequence (total 16S rDNA, *amo*, *Nitrospira*-16S rDNA) can be determined from the fluorescent measurements. The fluorescent threshold cycle (the cycle where the fluorescent signal is detected above background) is inversely proportional to the log of the initial copy number, i.e., the more template present at the beginning, the fewer cycles it takes to reach the fluorescent threshold (Higuchi et al., 1993).

New primer sets, specific for the *amo* genes from activated sludge and for the 16S rDNA of *Nitrobacter* sp. and *Nitrospira* sp., have been developed (Table 3-7). These primers were designed to produce DNA fragments <150 base pairs long. The primers 1392f and1525r, used to normalize total bacterial cell number in activated sludge (working unspecific; for all bacteria), were previously designed for amplification of a 16S rDNA fragment.

Table 3-7. Characteristics of Primers for Use in Molecular Beacon PCR.

Primer	Target gene	Sequence (5'-3')	T_M(˚C)
Amo267f	Ammonia monooxygenase gene	GAATATGTTCGCCTGATTG	58.0
Amo387r	Ammonia monooxygenase gene	CAAAGTACCACCATACGCAG	60.6
Ntspa1130f	*Nitrospira* sp.	CCTGCTTTCAGTTGCTACCG	62.0
Ntspa1260r	*Nitrospira* sp.	GTTTGCAGCGCTTTGTACCG	62.0
NB1290f	*Nitrobacter* sp.	GTGACAATGGGAAGCAAAGG	60.0
NB1420r	*Nitrobacter* sp.	CGTATTCACCGTGGCATGC	60.0
1392f	16S rDNA	GYACACACCGCCCGT	48.0
1525r	16S rDNA	AAGGAGGTGWTCCARCC	46.0

The primers for molecular beacon PCR were tested to verify that they amplify DNA fragments from activated sludge samples with the correct size and containing the correct DNA sequences. The correct size DNA fragment (121 bp) was amplified from plasmids containing cloned *amo* genes from the full-scale municipal WWTP activated sludge and bench-scale reactor activated sludge samples using the primer set *amo*267f and *amo*387r. These amplified DNA fragments were cloned into *E. coli* to check their DNA sequences. The plasmids from 10 *E. coli* colonies were isolated for DNA sequencing of the cloned PCR fragment. The primer sets for the amplification of *Nitrospira* and *Nitrobacter* 16S rDNA sequences were tested using the same protocols.

The molecular beacons for the quantification of the *amo* gene, the 16S rDNA of *Nitrospira* sp. and the universal 16S rDNA have been designed (Table 3-8). After synthesis, melting characteristics of the beacons were verified experimentally using the molecular beacon thermocycler.

Table 3-8. Characteristics of Beacons for Molecular Beacon PCR.

Molecular beacon	Target gene	Sequence (arm|beacon probe|arm; 5'-3')	T_M of stem (°C)	T_M of probe/target hybrid (°C)
Amobeac1	*amo*	GCCGT|GCAAGGCTCACTGCGTAC|ACGGC	66.9	63.0
Ntspabeac1	16S rDNA (*Nitrospira*)	GCCGCTCG|AGCACTCTGAAAGGACTGCCC AGG|CGAGCGGC	71.6	68.0
16Sbeac1	16S rDNA	GAAG|ACGGYTACCTTGTTACG|TCTTC	59.3	54.0
16Sbeac2	16S rDNA	GAGA|ACGGYTACCTTGTTACG|TTCTC	59.3	54.0

For quantification of the *amo*A gene by real-time PCR, the primers used in the competitive *amo*A PCR protocol (*amo*598f and *amo*718r primers) plus an additional molecular beacon probe internal to those primers were tested. The Amobeac1 molecular beacon was designed to bind to the *amo*A genes from the full-scale municipal WWTP. When additional *amo*A genes were obtained from the 2-day SRT bench-scale reactor, it was discovered that Amobeac1 would not bind efficiently to these sequences because of DNA sequence mismatches. Therefore, an additional beacon, Amobeac2 was designed to match the reactor sequences. For reasons not completely elucidated, neither of these molecular beacons was effective in quantifying the *amo*A gene in bench-scale biosolids samples.

3.3.1.2.7 TaqMan PCR

The TaqMan methodology differs from the molecular beacon methodology in that it is based on a linear probe with fluorophore and quencher, rather than a hairpin probe structure with the fluorophore and quencher. In the TaqMan methodology, fluorescence is detected during the PCR reaction by cleavage of the fluorescent molecule from the probe after it hybridizes to the target PCR product. Unlike the molecular beacon methodology, the fluorescence produced cannot be quenched by disassociation of the probe from the target and reformation of the hairpin probe structure.

Real-time PCR assays using both the molecular beacon and TaqMan probes for *Nitrospira* 16S rDNA were tested using the primers NSR1113f and NSR1264r and the probe sequences listed in Table 3-4. The initial application of real-time PCR using molecular beacon probes for the detection of *Nitrospira* was variable and produced inconsistent results. To determine whether a TaqMan probe would be more effective for quantifying *Nitrospira* than the molecular beacon, a TaqMan probe was designed using the same target sequence as the molecular beacon probe. In PCR reactions using *Nitrospira* 16S rDNA as the target, the TaqMan probe showed an eight-fold higher fluorescence and less variability than the molecular beacon probe. The TaqMan methodology accurately detected *Nitrospira* 16s rDNA (r^2=0.99) over six orders of magnitude, compared to only three orders of magnitude for the molecular beacon probe (r^2 = 0.90). In September 2000, the iCycler real-time PCR system was updated with new iCycler software (Version 2.1) and hardware for multiplex PCR. The new software and hardware enabled the system to determine a plate factor from the 96 well plate prior to each PCR run to normalize the background signals from each single well. The Platinum Taq Supermix was replaced by Platinum quantitative PCR Supermix-UDG (Gibco BRL) as the reagent mix in the PCR reactions. The use of the new PCR reagents with the upgraded iCycler system also resulted in improved signals from the molecular beacon Ntspabeac. However, TaqMan probes were selected for use in the collection of data in this research project because they were found to be more cost effective and easier to design than molecular beacon probes.

Based on new *amo*A sequence information from Purkhold et al. (2000) and our own libraries (Dionisi et al., 2001), broader sets of primers and TaqMan probes were designed for the detection of the *N. marina* type AOB in the municipal samples and for the *N. nitrosa* type AOB in the industrial samples (Table 3-4).

In addition to quantifying nitrifying bacteria, it is useful to quantify total bacteria so that data may be normalized with respect to biomass and DNA extraction efficiency. The standard method for determining total bacterial 16S rDNA uses dot blot hybridizations with a ^{32}P-labeled oligonucleotide. However, dot blot hybridizations are laborious and time consuming (3 days per experiment). Therefore, methods were developed to quantify total 16S rDNA using real-time PCR. A Taqman probe (16S TAQ 1114f,) was designed for the detection the of total 16S rDNA by real-time PCR using the primers 1055f and 1392r (Lane, 1991). For these reactions, Platinum quantitative PCR Supermix-UDG (GibcoBRL) with 600 nM primers, 250 nM 16S TAQ 1114f, and 0.5 ng to 1 ng template DNA was used. The amplification program consisted of an initial denaturation at 95°C for 10 minutes; followed by 11 cycles of denaturation at 95°C for 30 seconds; annealing at 57°C for 60 seconds, decreasing 1°C per cycle, and extension at 72°C for 20 seconds; followed by 29 cycles of denaturation at 95°C for 30 seconds, annealing at 47°C for 60 seconds, and extension at 72°C. Fluorescence was detected during the annealing step of each cycle of the PCR run.

Quantification of 16S rDNA was linear ($r^2 = 0.99$) over four orders of magnitude. This range was lower than that obtained with the *Nitrospira* Taqman system due to the type of PCR program used and 16S rDNA contamination of the Taq polymerase. The PCR program used for amplification of total 16S rDNA is a touchdown program, where data is not acquired in the first 12 PCR cycles. If there are too many copies of 16S rDNA initially in the reaction tube, then the threshold cycle is located within the first 12 cycles and the baseline cannot be defined. Therefore, the samples need to be diluted to below 10^6 16S rDNA copies per reaction. The 16S rDNA contamination in Taq polymerase results from the recombinant *E. coli* cells used to produce the Taq enzyme and has been previously reported (Corless et al., 2000, Suzuki et al., 2000). In these experiments, the background levels of 16S rDNA from the Taq enzyme were 3×10^3 copies per PCR reaction. However, because the total 16S rDNA in undiluted DNA extracts is typically 10^8 copies per PCR reaction, neither of these issues resulted in significant quantification problems.

3.3.2 Real-time TaqMan PCR Assays

Real-time PCR assays were developed and optimized for the quantification of bacterial 16S rDNA, *Nitrospira* spp. 16S rDNA, *N. oligotropha*-type *amo*A, and *N. nitrosa amo*A in mixed liquor samples. Four TaqMan probes, 16Staq 1115, *amo*NoTaq 729, *amo*NnTaq 740, and NSR 1143-Taq (Table 3-9), were designed using the guidelines provided by Applied Biosystems (http:/home.appliedbiosystems.com; Applied Biosystems, Foster City, CA). The primers and probes were synthesized by Sigma Genosys (Sigma Genosys, The Woodlands, TX) (Harms et al., 2003).

Table 3-9. Primers and Probes Used in this Study.

Assay	Target	Primer/Probe[a]	Sequence (5'-3')[b]	Reference
Bacterial 16S rDNA	Bacteria 16S rDNA	1055f	5'-ATGGCTGTCGTCAGCT-3'	Lane, 1991
		1392r	5'-ACGGGCGGTGTGTAC-3'	Thompson et al., 1994
		16STaq1115f	5'-(6-FAM)-CAACGAGCGCAACCC-(TAMRA)-3'	Harms et al., 2003
N. oligotropha-type *amoA*	*Nitrosomonas oligotropha amoA* gene	*amo*No550D2f	5'-TCAGTAGCYGACTACACMGG-3'	Harms et al., 2003
		*amo*No754r	5'-CTTTAACATAGTAGAAAGCGG-3'	Harms et al., 2003
		*amo*NoTaq729r	5'-(6-FAM)-CCAAAGTACCACCATACGCAG-(TAMRA)-3'	Harms et al., 2003
N. nitrosa amoA	*Nitrosomonas nitrosa amoA* gene	*Amo*Nn549f	5'-TTTCAATGGCAGACTACA-3'	This study
		*amo*Nn777r	5'-TTTAACAATACGGCCTCT-3'	This study
		*amo*NnTaq740	5'-(6-FAM)-AATGCTGTACAGTAAACTTTACCGAG-(TAMRA)-3'	This study
Nitrospira 16S rDNA	*Nitrospira* spp. 16S rDNA	NSR1113f	5'-CCTGCTTTCAGTTGCTACCG-3'	Johnsen et al., 1999
		NSR1264r	5'-GTTTGCAGCGCTTTGTACCG-3'	Johnsen et al., 1999
		NSR1143fTaq	5'-(6-FAM)-AGCACTCTGAAAGGACTGCCCAGG-(TAMRA)-3'	Harms et al., 2003

[a] Primer/Probe abbreviations: f= forward primer, r= reverse primer, Taq= TaqMan probe.
[b] 5'-fluorophore-probe-quencher-3' in case of TaqMan probe. 6-FAM = 6-carboxyfluorescein; TAMRA = carboxytetramethylrhodamine.

3.3.2.1 Development of Real-time PCR for Quantification of Bacterial 16S rDNA

Amplification of total bacterial 16S rDNA was performed using primers 1055f (Ferris et al., 1996) and 1392r (Lane et al., 1991). The TaqMan probe 16STaq1115f was modified from the 1114f primer (Lane et al., 1991). The composition of the PCR mix with a total volume of 25 µl is shown in Table 3-10. Dilutions of plasmid pCR®2.1 vector (Invitrogen, Carlsbad, CA) carrying an uncultured bacterium 16S rRNA gene for *Nitrospira* (GenBank accession number

Table 3-10. PCR Mix for the Quantification of Total 16S rDNA.

Ingredient	Volume in one 25 μl reaction (in μl)
Platinum® Quantitative PCR SuperMix-UDG (Life Technologies, Inc., Gaithersburg, MD)	12.5
Primer 1055f (20 μM stock)	0.75
Primer 1392r (20 μM stock)	0.75
$MgCl_2$ (50 mM stock)	1.0
16STaq1115f TaqMan probe (10 μM stock)	0.625
sterile HPLC water	4.375
Sample DNA or dilutions of standard	5.0

AF420301) (Dionisi et al., 2002) ranging from 4.5×10^3 to 4.5×10^8 copies of the 16S rDNA gene were used as standards in this assay. The PCR program is depicted in Figure 3-7 and consisted of 3 minutes at 50°C, an initial denaturation at 95°C for 10 minutes, followed by 45 cycles of denaturation at 95°C for 30 seconds, annealing at 50°C for 60 seconds, and extension at 72°C. Fluorescence was detected during the annealing step of each cycle of the PCR run.

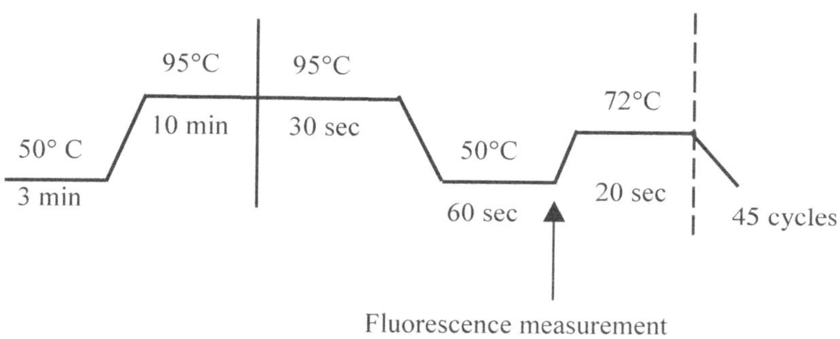

Figure 3-7. PCR Program for the Quantification of Total 16S rDNA.

3.3.2.2 Development of Real-time PCR for Quantification of *N. oligotropha*-like *amoA* Gene

The primers *amo*No550D2f and *amo*No754r were designed to target the *amo*A gene of AOB found in the full-scale municipal WWTP (Dionisi et al., 2002) based on alignment of *amo*A gene sequences using the CLUSTAL W program (Purkhold et al., 2000). Alignments consisted of *amo*A sequences from clonal libraries obtained from the WWTP (Johnsen et al., 1999), four bench-scale municipal wastewater treatment systems (Dionisi et al., 2002), and *amo*A sequences available in GenBank (Chandler, 1998). The forward primer (*amo*No550D2f) contained two degenerate bases in order to amplify all *amo*A clones from the libraries, as well as *N. urea* (AJ388585) and *N. oligotropha* (AF272406) *amo*A genes. The TaqMan probe *amo*NoTaq729r was derived from a conserved sequence region within the primer pair *amo*No550D2f and *amo*No754r. The composition of the PCR mix with a total volume of 25 μl is shown in Table 3-11. Standards consisted of the plasmid pCR®2.1 carrying the M-20 *amoA* gene

(GenBank accession number AF420299) (Dionisi et al., 2002) adjusted to 10 to 1.0 × 10^7 copies per PCR reaction. The PCR amplification program is depicted in Figure 3-8 and consisted of 3 minutes at 50°C, 10 minutes at 95°C, 55 cycles at 95°C for 30 seconds, and 56°C for 60 seconds.

Table 3-11. PCR Mix for the Quantification of *Nitrosomonas oligotropha*-type *amo*A.

Ingredient	Volume in one 25 μl reaction (μl)
TaqMan® Universal PCR Master Mix (PE Applied Biosystems, Foster City, CA)	12.5
Primer amoNo550D2f (20 μM stock)	0.375
Primer amoNo754r (20 μM stock)	0.375
TaqMan probe *amo*NoTaq729r (10 μM stock)	0.625
Sterile HPLC water	6.125

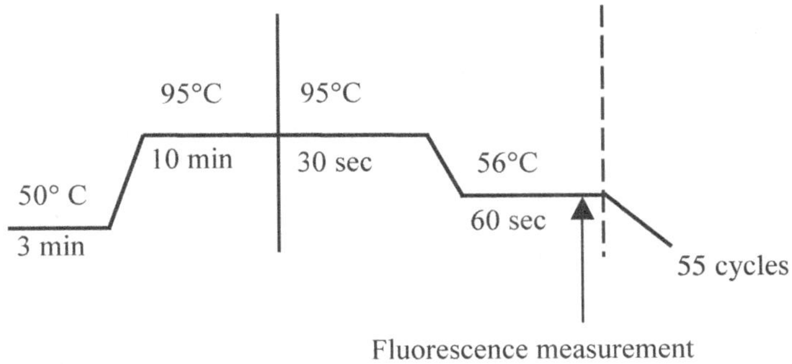

Figure 3-8. PCR Program for the Quantification of *Nitrosomonas oligotropha*-type *amo*A.

3.3.2.3 Development of Real-time PCR for Quantification of *N. nitrosa amo*A Gene

Previous studies of the ammonia-oxidizing population in the industrial full-scale WWTP using *amo*A suggested the presence of a *Nitrosomonas nitrosa*-like AOB in the suspended solids, which belong to the *N. europaea* cluster (Dionisi et al., 2002). Sequences of the *amo*A gene retrieved from library and pure culture data were used to design a set of primers and a probe (Table 3-9) for the quantification of *N. nitrosa* in suspended solids using real-time PCR. The real-time PCR assay amplified a 230 bp sequence of the *amo*A gene from *N. nitrosa* (bases 200 to 429, NCBI Accession Number AF272404). The PCR reaction for this set of primers and probe was optimized. The optimal conditions for the real-time PCR reaction were: Ready-To-Go PCR beads (Amersham Pharmacia, Piscataway, NJ), 5 mM $MgCl_2$, 100 nM forward and reverse primers, and 100 nM probe. The program used for amplification was 95°C for 10 minutes, then 55 cycles consisting of 95°C for 15 seconds and 54°C for 30 seconds. A no-template control was included in each run. Amplification and detection were undertaken using the DNA Engine Opticon™ Continuous fluorescence Detection System (MJ Research, Waltham, MA) or a Smart Cycler System (Fisher Scientific, Pittsburgh, PA). The threshold value for all measurements was chosen to be 0.01 for the DNA Engine Opticon™ System, and 10 in the Smart Cycler System, both within the range of logarithmic fluorescence increase.

The specificity of the primer set was tested by PCR amplification of industrial WWTP genomic DNA samples in the same conditions used for the real-time PCR reaction, followed by cloning and sequencing of 14 clones from the library. All analyzed clones were found to be *amo*A and presented between 99.1% and 100% identity with the sequence reported for *N. nitrosa* and the *amo*A sequence of E1 strain.

3.3.2.4 Development of Real-time PCR for Quantification of *Nitrospira* 16S rDNA

The *Nitrospira* 16S rDNA primers NSR1113f and NSR1264r were deduced from 16S rDNA sequences aligned with the CLUSTAL W program (Thompson et al., 1994). The alignment contained two *Nitropsira* partial 16S rDNA sequences obtained from industrial MLSS (Cook et al., 2000) and *Nitrospira* 16S rDNA sequences available from GenBank (National Center for Biotechnology Information). The *Nitrospira* clones from the industrial MLSS, 931 (AF420301) and 9335, were 99% similar to each other and 98% similar to other *Nitrospira* 16S rDNA sequences including Y14639, AF155154, and AJ224038.

The TaqMan probe NSR1143fTaq was derived from a conserved sequence region between the primers NSR1113f and NSR1264r. Real-time PCR assays using NSR1143fTaq were performed in a total volume of 25 μl with the composition of the PCR mix shown in Table 3-12.

Table 3-12. PCR Mix for the Quantification of *Nitrospira* 16S rDNA.

Ingredient	Volume in one 25 μl reaction (in μl)
Platinum® Quantitative PCR SuperMix-UDG (Life Technologies, Inc., Gaithersburg, MD)	12.5
Primer NSR1113f (20 μM stock)	0.75
Primer NSR1264r (20 μM stock)	0.75
$MgCl_2$ (50 mM stock)	1.0
NSR1143fTaq (10 μM stock)	0.625
sterile HPLC water	4.375

Standards consisted of 151 bp fragment of *Nitrospira* 16S rDNA from (Dionisi et al., 2002) adjusted to 30 to 3×10^7 copies. The PCR amplification (Figure 3-9) consisted of 3 minutes at 50°C, 10 minutes at 95°C, 55 cycles at 95°C for 30 seconds, and 63°C for 60 seconds.

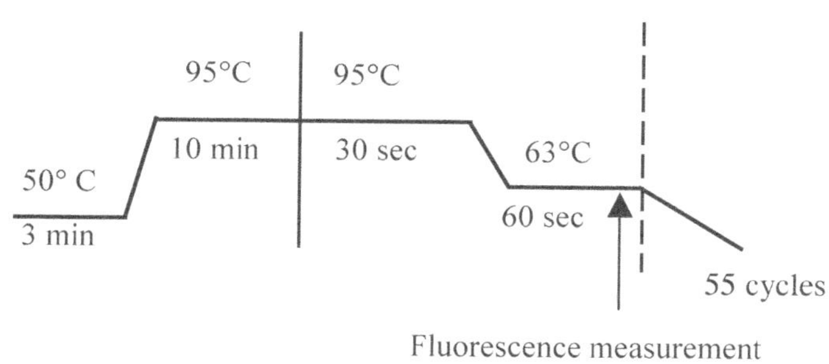

Figure 3-9. PCR Program for the Quantification of *Nitrospira* 16S rDNA.

3.3.2.5 Acquisition and Data Analysis

Real-time PCR assays for bacterial 16S rDNA and *Nitrospira* 16S rDNA were run on a Bio-Rad iCycler with the iCycler iQ™ fluorescence detector and iCycler software Version 2.3 (Bio-Rad, Hercules, CA). Plate well factors were determined prior to each PCR run to normalize background fluorescence intensities from each single well. *Amo*A real-time PCR assays were run using a DNA Engine Opticon™ Continuous Fluorescence Detection System (MJ Research, Waltham, MA). The threshold was determined by the computer software as 10 times the standard deviation of the background fluorescence averaged over at least five cycles at the start of the run. The threshold cycle (C_T) of each PCR reaction was automatically determined by detecting the cycle at which the fluorescence exceeded the calculated threshold. During each PCR run, the C_T values obtained from the DNA standards were used for the construction of standard curves.

3.3.2.6 Application of Real-time PCR Assays to MLSS Samples

All real-time PCR assays were performed using three replicates per sample, and all PCR runs included control reactions without template. The effect of sample concentration on PCR performance was determined using dilutions of sample DNA (initial concentration of 50 ng/µl) containing 5 ng to 50 ng in sterile water followed by real-time PCR analysis for 16S rDNA, *Nitrospira* 16S rDNA, and *N. oligotropha*-like *amo*A as described above.

Gene copies were initially calculated by comparison of threshold cycles obtained in each PCR run from known standard DNA concentrations. To reduce variability between PCR runs, data were recalculated using a second standard curve generated from 11 standard curves for bacterial 16S rDNA ($r^2 = 0.94$), 22 standard curves for *Nitrospira* 16S rDNA ($r^2 = 0.84$), and five standard curves for *N. oligotropha*-like *amoA* ($r^2 = 0.90$). In the case of the *Nitrospira* 16S rDNA and bacterial 16S rDNA assays, one universal standard curve was applied for calculations, because both assays were shown to function alike with standard plasmid AF420301 as template.

3.3.3 Statistical Analysis

Multiple linear regression analyses were performed on the physical/chemical and molecular data generated in this study using the SPSS 11.0 (SPSS, Inc., Chicago, IL) statistical analysis software package. This type of analysis was used to estimate models to describe the distribution of a response (or independent) variable with a number of predictors (or dependent variables). A function of the analysis was to search for predictor variables that explain significant variation in the response variable. If a number of significant predictors could be identified, the model provided information as to the relative influence each parameter had on the response variable.

In this study, the predictor variables were the controlled parameters, SRT, temperature and DO concentration. The response variables were the measured parameters MLVSS, total eubacteria concentration, ammonia and nitrite oxidation rates, and the *N. oligotropha*-type AOB and *Nitrospira* NOB concentrations. In addition to these main effect predictor variables, interaction terms were included in the models to incorporate the joint effect of two variables (e.g., temperature and DO) on a dependent variable (e.g., ammonia oxidation rate) over and above their separate effects. The two-way interactions of SRT and temperature, SRT and DO, and temperature and DO, as well as the three-way interaction of SRT, temperature, and DO were included in the multiple linear regression analyses. The interaction terms were added to the models as cross products of the individual independent variables. When studying the measures of biomass (MLVSS and total eubacteria concentration), the influent COD level was also considered in the linear regression models since this parameter did vary with each batch of influent wastewater. When studying nitrification, the influent ammonia level was incorporated into the analysis.

In order to perform multiple linear regression analyses, it was necessary to assume that the relationship between the dependent and the independent variables was linear and that for each combination of independent variable values the distribution of the dependent variable was normal with a constant variance (Norušis, 2000). To meet these requirements, it was necessary to first transform some of the data. The logarithm of the molecular data (total eubacteria, *N. oligotropha*-type AOB, and *Nitrospira* concentrations) had to be computed before inputting this data into the analyses.

Multiple linear regression fits a response variable as a linear combination of multiple X variables by the method of least squares analysis. This hypothetical relationship is described in the following equation:

$$Y = C + B_1X_1 + B_2X_2 + \ldots B_kX_k \qquad (3.3.3\text{-}1)$$

where:

Y = independent (response) variable
C = constant
B_1, B_2, B_k = partial regression coefficients of predictors
X_1, X_2, X_k = dependent (predictor) variables

In the multiple linear regression model, the constant (C) is determined by the intercept of the regression line with the y-axis. This number represents the value of Y when all independent variables are zero. The partial regression coefficient (B) of a predictor quantifies the amount of linear trend in Y. It is the average amount the dependent variable increases when the independent parameter increases one unit and other independents are held constant (Norušis, 2000). In other words, the B coefficient is the slope of the regression line. The larger the value of B, the steeper the slope, and the more the dependent variable changes for each unit change in the independent variable. A positive coefficient means that the predicted value of the dependent variable increases when the value of the dependent variable increases. Conversely, a negative coefficient means that the predicted value of the dependent variable decreases when the value of the independent parameter increases.

The null hypothesis, that the partial regression coefficient for a variable is zero, was tested using the t-statistic and its observed significance level (Norušis, 2000). If the confidence interval for the partial regression coefficient includes 0, then there is no significant linear relationship between X and Y, and the null hypothesis cannot be rejected. A partial regression coefficient of zero would indicate that the predictor had no effect on the response variable.

Standardized partial regression (beta) coefficients can be compared to judge the relative importance of each independent variable in the model. These coefficients enable the comparison of variables of differing magnitudes and measurement units and are equal to the average amount the dependent variable increases when the independent variable increases one standard deviation and the other independent variables are held constant (Norušis, 2000).

Before generalizations can be made about the experimental data based on the results of the multiple linear regression analysis, it is necessary to confirm how well the model fits the data. One measure of goodness of fit is the coefficient of multiple determination, R^2. This parameter indicates how much of the observed variability in the dependent variable can be explained collectively by all of the independent variables (Norušis, 2000). An R^2 value close to 1.0 indicates that almost all of the variability has been accounted for with the variables specified in the model. The degree to which two or more predictors are related to the dependent variable is expressed in the correlation coefficient R, which is the square root of R^2. R is the correlation coefficient between the observed value and the predicted value based on the regression model. A value of 1 indicates the dependent variable can be perfectly predicted from the independent variables. A value close to zero indicates that the independent variables are not linearly related to the depended variable.

Analysis of variance (ANOVA) was used to test several equivalent null hypotheses. The first is that there is no linear relationship to the population between the dependent and independent variables. The second is that all of the partial regression coefficients are zero. The last is that the population value for multiple R^2 is zero. The test of the null hypotheses is based on the ratio of the regression mean square to the residual mean square and is called the overall regression F-test (Norušis, 2000). The F-test is used to test the significance of R, which is the same as testing the significance of the model as a whole. If the observed significance level of F is <0.05, then the model is considered significantly better than would be expected by chance and the null hypothesis, that no linear relationship exists between the dependent and independent variables, can be rejected.

CHAPTER 4.0

RESULTS AND DISCUSSION

4.1 Municipal Bench-scale Treatment System

Four activated sludge reactors treating municipal wastewater were operated at three constant temperature levels (20°C, 15°C, and 10°C) and varying DO levels. The purpose was to assess the impact of these parameters on the effectiveness of activated sludge performance. The study of DO effects on treatment performance at 20°C was conducted between June 2000 and January 2001. The DO was initially set at 4.0 mg/L to minimize the impact of limited DO on nitrogen treatment performance. It was lowered in discrete increments to a final value of 0.5 mg/L for the 2-day SRT reactor and 0.2 mg/L for the 5-, 10-, and 20-day SRT reactors. The study at 15°C took place between March 2001 and September 2001. The DO concentration was set at 3.0 mg/L to begin the experiment and was lowered in discrete increments to a final DO level of 0.2 mg/L. The 10°C study was conducted between October 2001 and January 2002. DO concentrations assessed included 2.0 mg/L, 1.0 mg/L, and 0.5 mg/L. For analysis purposes, reactor DO concentrations were grouped into high (≥2.0 mg/L), intermediate (1.0 mg/L to 1.5 mg/L) and low (≤0.5 mg/L) levels. The time frames for operation of each reactor at various DO levels are reported in Tables 4-1, 4-2, and 4-3 for 20°C, 15°C, and 10°C temperatures, respectively.

4.1.1 COD Treatment Performance

Nitrification performance was of primary interest in this study. However, because nitrifying bacteria in single-stage nitrification systems are indirectly affected by the presence of heterotrophic organisms, the activity of these organisms was also monitored. Carbon treatment performance was used to indicate heterotrophic activity in the reactors and was documented by performing COD analyses of the influent and reactor effluents. While COD_T was used to monitor influent strength, effluent samples were filtered to minimize the influence of suspended solids on effluent COD values. The influent wastewater was originally filtered as well. However, a large fraction of the COD was removed during this process; consequently, filtration of influent wastewater was not performed during this study. Therefore, carbon treatment performance measures reported here are indicative of each treatment unit's effectiveness in removing soluble, degradable COD from the influent wastewater.

4.1.1.1 20°C Operation

COD removal efficiency was used as a measure of carbon treatment performance and was computed according to the following equation:

$$\%\,\text{COD Removal} = \frac{\text{COD}^0 - \text{COD}^e}{\text{COD}^0} \times 100\% \qquad (4.1.1.1\text{-}1)$$

where:

COD^0 = influent COD concentration (mg/L)
COD^e = effluent COD concentration (mg/L)

Figure 4-1 shows influent and effluent COD levels over the 20°C study. Table 4-4 summarizes carbon treatment performance for the four reactors during the high, intermediate, and low DO periods. COD removal was excellent in all four reactors during the 20°C operational period, with the exception of the week of July 28 to August 4, 2000. During this week, the presence of what is believed to be a surfactant-type material decreased the degradability of the influent and caused high effluent COD values for all SRTs. Once the surfactant passed through the system, effective COD treatment resumed. The full-scale municipal WWTP also had similar problems with this unknown material and exceeded the permitted maximum daily BOD discharge level four times during that week.

Table 4-1. DO levels and Operational Dates for the 20°C Municipal Study.

DO Level	High	Intermediate	Low
20-Day SRT	6/6/00-9/6/00	9/7/00-11/2/00	11/3/00-1/10/01
10-Day SRT	6/6/00-9/6/00	9/7/00-11/2/00	11/3/00-1/10/01
5-Day SRT	6/6/00-9/6/00	9/7/00-11/2/00	11/3/00-1/10/01
2-Day SRT	6/6/00-10/27/00	10/28/00-12/27/00	12/28/00-1/10/01

Table 4-2. DO levels and Operational Dates for the 15°C Municipal Study.

DO Level	High	Intermediate	Low
20-Day SRT	3/1/01-4/25/01	4/26/01-8/1/01	8/2/01-9/29/01
10-Day SRT	3/1/01-4/25/01	4/26/01-8/1/01	8/2/01-9/29/01
5-Day SRT	3/1/01-4/25/01	4/26/01-8/1/01	8/2/01-9/29/01
2-Day SRT	3/1/01-4/25/01	4/26/01-8/1/01	8/2/01-9/29/01

Table 4-3. DO levels and Operational Dates for the 10°C Municipal Study.

DO Level	High	Intermediate	Low
20-Day SRT	10/23/01-11/29/01	11/30/01-12/12/01	12/13/01-1/6/02
10-Day SRT	10/23/01-11/29/01	11/30/01-12/12/01	12/13/01-1/6/02
5-Day SRT	10/23/01-11/29/01	11/30/01-12/12/01	12/13/01-1/6/02
2-Day SRT	10/23/01-11/29/01	11/30/01-12/12/01	12/13/01-1/6/02

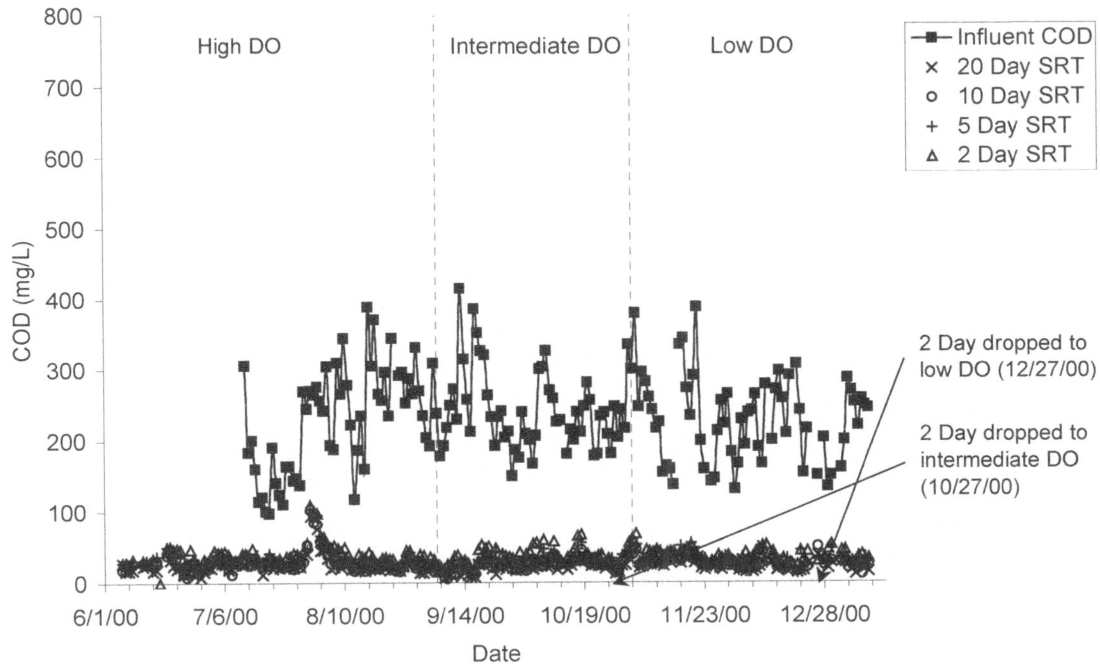

Figure 4-1. Effect of DO Concentration and SRT on COD Treatment Performance During 20°C Operation.

Table 4-4. Average (±SD) COD Treatment Performance During Operation at 20°C.

DO Level	High		Intermediate		Low	
	Avg. COD (mg/L)	% COD Removal	Avg. COD (mg/L)	% COD Removal	Avg. COD (mg/L)	% COD Removal
Influent	230 (±74)	n/a	242 (±56)	n/a	229 (±61)	n/a
20-Day SRT	26 (±14)	87 (±7)	22 (±7)	91 (±3)	25 (±8)	88 (±5)
10-Day SRT	29 (±15)	86 (±7)	25 (±8)	89 (±4)	27 (±7)	87 (±5)
5-Day SRT	32 (±16)	83 (±9)	31 (±10)	87 (±5)	33 (±10)	85 (±6)
2-Day SRT	38 (±15)	82 (±7)	38 (±10)	83 (±5)	23 (±8)	81 (±9)

The 20-day SRT reactor consistently provided the highest average treatment efficiency under high, intermediate, and low DO conditions. As expected, the 2-day SRT reactor provided somewhat poorer average treatment. However, COD removal remained above 80% in this reactor and did not vary substantially with changes in DO. This was not surprising since the oxygen half-saturation constant for heterotrophic bacteria is on the order of a few tenths of a microgram per liter (Rittmann and McCarty, 2001).

4.1.1.2 15°C Operation

Influent and effluent COD levels are plotted over time for the 15°C municipal study (Figure 4-2). Table 4-5 summarizes the average carbon treatment performance at each operational phase. In general, average treatment performance decreased with decreasing SRT, but was unaffected by decreasing DO conditions. The 15°C results were similar to the 20°C results.

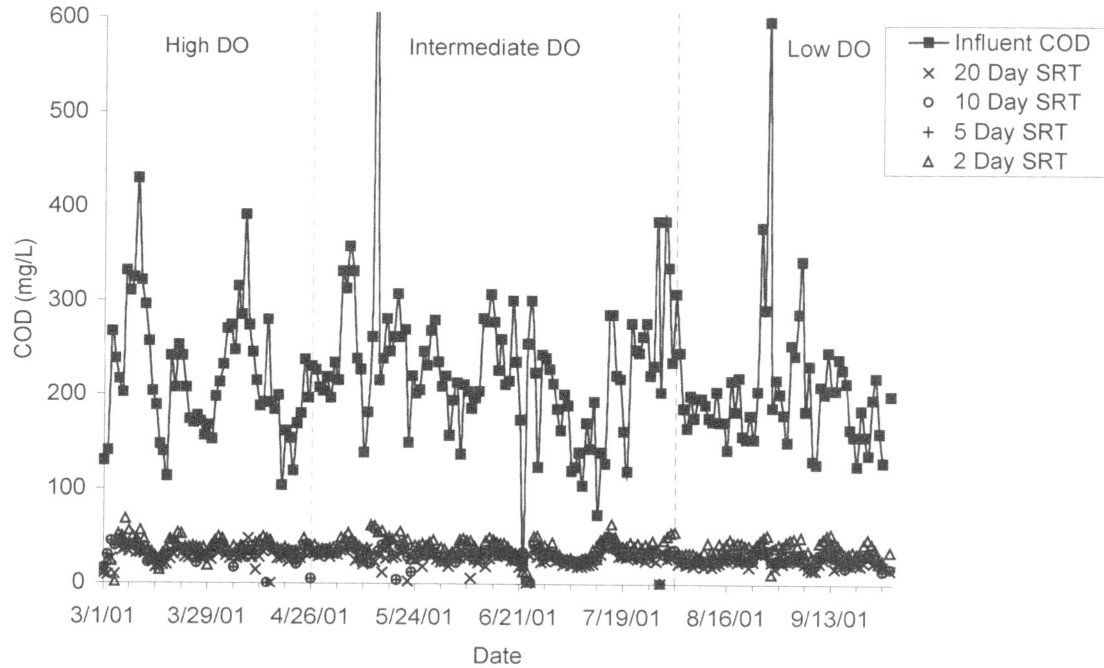

Figure 4-2. Effect of DO Concentration and SRT on COD Treatment Performance During 15°C Operation.

Table 4-5. Average (±SD) COD Treatment Performance During Operation at 15°C.

DO Level	High		Intermediate		Low	
	Avg. COD (mg/L)	% COD Removal	Avg. COD (mg/L)	% COD Removal	Avg. COD (mg/L)	% COD Removal
Influent	219 (±68)	n/a	227 (±82)	n/a	204 (±72)	n/a
20-Day SRT	27 (±8)	87 (±5)	27 (±7)	86 (±13)	23 (±5)	88 (±3)
10-Day SRT	33 (±8)	84 (±5)	31 (±9)	84 (±17)	27 (±6)	86 (±3)
5-Day SRT	36 (±10)	82 (±5)	38 (±10)	81 (±18)	32 (±7)	84 (±4)
2-Day SRT	39 (±11)	81 (±6)	39 (±11)	81 (±8)	38 (±8)	80 (±5)

4.1.1.3 10°C Operation

Figure 4-3 shows influent and effluent COD levels throughout the 10°C study period, while Table 4-6 depicts average COD treatment performance for each DO level. Figure 4-3 reveals that the strength of the influent wastewater declined throughout the course of the 10°C study, so much so that the average influent COD during the low DO phase was only

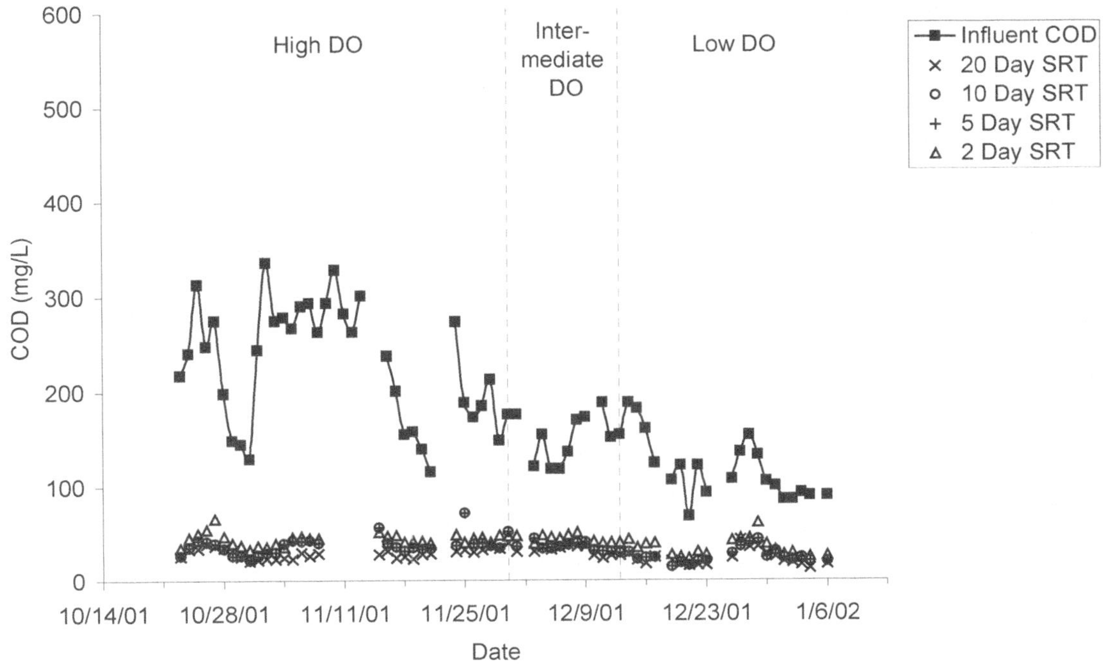

Figure 4-3. Effect of DO Concentration and SRT on COD Treatment Performance During 10°C Operation.

Table 4-6. Average (±SD) COD Treatment Performance During Operation at 10°C.

DO Level	High		Intermediate		Low	
	COD (mg/L)	% Removal	COD (mg/L)	% Removal	COD (mg/L)	% Removal
Influent	230 (±63)	n/a	154 (±26)	n/a	120 (±33)	n/a
20-Day SRT	29 (±5)	88 (±7)	33 (±5)	78 (±5)	23 (±8)	80 (±5)
10-Day SRT	37 (±10)	85 (±8)	37 (±6)	75 (±6)	26 (±8)	78 (±6)
5-Day SRT	40 (±8)	83 (±9)	40 (±4)	73 (±7)	30 (±7)	74 (±6)
2-Day SRT	44 (±7)	82 (±9)	45 (±4)	70 (±6)	35 (±10)	70 (±6)

one-half of that during the high DO phase. Tables 4-5 and 4-6 show that average percent COD removal decreased at all SRTs for the intermediate and low DO levels when the temperature was decreased from 15°C to 10°C. However, this trend can be attributed to the decrease in influent COD values, rather than a decline in treatment performance. The impact of influent COD concentration on percent COD removal can be shown by rearranging equation 4.1.1.1-1 as follows:

$$\% \text{ COD Removal} = \left(\frac{COD^0}{COD^0} - \frac{COD^e}{COD^0} \right) \times 100\% = \left(1 - \frac{COD^e}{COD^0} \right) x\, 100\% \qquad (4.1.1.3\text{-}1)$$

Equation 4.1.1.3-1 indicates that for a constant COD^e value, a decrease in COD^0 would yield a larger value for the ratio of COD^e/COD^0, which when subtracted from 1 would result in a lower reported COD removal efficiency. As shown in Table 4-6, average effluent COD levels during intermediate and low DO operation at 10°C were similar to values measured at high DO for this temperature. Therefore, while the effluent COD concentrations were stable during the latter stages of this experiment, the large decrease in influent COD levels could result in lower COD removal efficiencies according to equation 4.1.1.3-1. Average effluent COD concentrations during the low DO phase are lower than those observed during the intermediate and high DO phases at this temperature, suggesting that DO did not negatively impact carbon treatment performance. Additionally, the average residual COD concentrations measured during 10°C operation at the low DO level are similar to those measured during 20°C and 15°C operation at the low DO level, indicating that temperature did not significantly affect carbon oxidation.

4.1.1.4 Overall Carbon Treatment Performance

Efficient COD removal was achieved throughout all phases of the experiment except for the minor process upset mentioned above that occurred from July 28 to August 5, 2000. Excluding this time period, average effluent COD levels remained below 50 mg/L for all SRTs. A large portion of this residual COD is most likely comprised of soluble organic materials that are resistant to biological breakdown (i.e., refractory organics). Davis and Cornwell (1998) reported that refractory organics can result in effluent-soluble COD values in the range of 30 mg/L to 60 mg/L for full-scale municipal WWTPs. These biologically resistant materials do not impact the 5-day biochemical oxygen demand (BOD_5) measurement, since BOD_5 only detects the oxygen requirement of biologically oxidizable organic materials. As a result, the COD of an effluent sample will generally be greater than the BOD_5 for the same sample. Currently in the United States, effluent requirements are stipulated in terms of BOD_5 rather than COD. In Tennessee, municipal and domestic WWTPs are required to achieve maximum average monthly discharges of 30 mg/L BOD_5 (TDPH, 1999). Conversion of this value into units of COD assuming the COD_T: BOD_5 ratio of 2.1:1 suggested by Grady et al. (1999) yields a value of 63 mg/L COD. Since the concentration of COD discharged from the reactors remained below 50 mg/L throughout all operating periods, the results of this study indicate that the bench-scale system could meet the carbon treatment goals of a full-scale WWTP.

While a high degree of carbon oxidation occurred even at the 2-day SRT, average COD removal efficiency increased with increasing SRT. This trend was expected based on equation 2.3-2, which shows the relationship between SRT and effluent substrate concentration, and is in agreement with the results of other studies reported in the literature. For example, when operating a pilot-scale activated sludge system to treat domestic wastewater with an average influent COD of 288 mg/L, McClintock et al. (1993) noted COD removal efficiencies of 88%, 89%, and 93% for SRTs of 1.2, 2.7, and 5 days, respectively. Urbain et al. (1998) obtained similar results in a pilot-scale membrane bioreactor. When operated at SRTs of 5, 10, and 20 days, average effluent COD concentrations were 19 mg/L, 18 mg/L, and 17 mg/L, respectively, resulting in COD removal efficiencies of 94%, 94%, and 96%, respectively. These results, which show that effluent quality improved with increasing SRT, are consistent with this study.

DO concentration did not appear to affect carbon removal efficiency in the reactors, which was not surprising since, as discussed in Section 2.5.1, the oxygen half-saturation constant for heterotrophic bacteria is in the range of 0.01 mg/L to 0.15 mg/L (Henze et al., 1987). Several studies have reported effective carbon treatment at low DO levels. Chuang et al. (1997) found that at SRTs of 5, 10, and 15 days, carbon treatment performance was not impacted by changes in DO concentrations between 0.1 mg/L and 2.0 mg/L in a combined activated sludge-biofilm process. For example, when the system was operated at a 10-day SRT and used to treat a synthetic feed with an influent COD concentration of 300 mg/L, COD removal efficiencies of 96.7%, 96.3%, and 96.3% were achieved at DO concentrations of 2.0, 0.5, and 0.1 mg/L DO, respectively. Similar findings were reported by Lau et al. (1984), who found that effective carbon removal could be achieved at DO concentrations as low as 0.088 mg/L. When operated at an SRT of 0.308 days and DO concentrations of 6.16 mg/L, 0.352 mg/L, and 0.088 mg/L, 98.0%, 96.3%, and 97.2% COD removal, respectively, were achieved in a chemostat treating synthetic wastewater.

Based on residual COD values for each of the treatment units, changes in temperature did not significantly affect COD levels discharged under the range of conditions studied. These results are consistent with empirical evidence indicating the process of carbon oxidation by heterotrophic bacteria is fairly insensitive to temperature because of the high solids levels maintained in activated sludge systems (Benefield and Randall, 1980). In a set of bench-scale activated sludge experiments, Sayigh and Malina (1978) found that, at SRTs of three days or more, BOD removal efficiency was independent of temperature variations between 4°C and 31°C. McClintock et al. (1993) obtained similar results in their investigation of the effects of temperature and SRT on biological nutrient removal (BNR). In their study, COD removal efficiencies of 88%, 91% and 87% were achieved at 20°C, 15°C, and 10°C, respectively, in a pilot-scale activated sludge system operated at an SRT of 15 days. Lishman et al. (2000) also found that residual COD concentrations were not significantly impacted by temperature changes in a sequencing batch reactor (SBR) operated at a 10-day SRT. When treating a synthetic feed with an influent COD level of 400 mg/L, effluent COD values ranged between 25 and 50 mg/L and did not vary substantially when the operating temperature was lowered from 19°C to 14°C.

4.1.2 Solids Analysis

Daily solids analysis was conducted to ensure that the reactors were operating at the target SRTs. Since settled wastewater (primary clarifier effluent) from the full-scale municipal WWTP was used as the influent feed to the bench-scale systems, waste characteristics (i.e. COD, nitrogen content, etc.) varied with each date of collection. Although SRT levels were maintained at fairly stable levels, strict steady state conditions were difficult to maintain due to the variability of the influent waste stream.

4.1.2.1 20°C Operation

For the 20°C operational period, a number of changes in MVLSS concentration can be noted over time (Figure 4-4). The first major change in solids concentration occurred on August 27th. This abrupt drop in MLVSS concentration was due to a tubing change in the influent pump. A thick walled type of tubing was used to replace worn tubing; however, the tubing was too thick and did not allow the full flow of influent wastewater to pass into the reactors. The 20-day SRT reactor was most affected by the change in tubing, dropping from a MLVSS concentration of ≈2310 mg/L to ≈1310 mg/L in three days. Solids in the 10-day SRT reactor decreased from ≈1460 mg/L to ≈1040 mg/L while the 2- and 5-day SRT reactors were only marginally affected. Once this problem was realized, the tubing was again replaced. Installation of the proper tubing resulted in an increased MLVSS level in all four reactors.

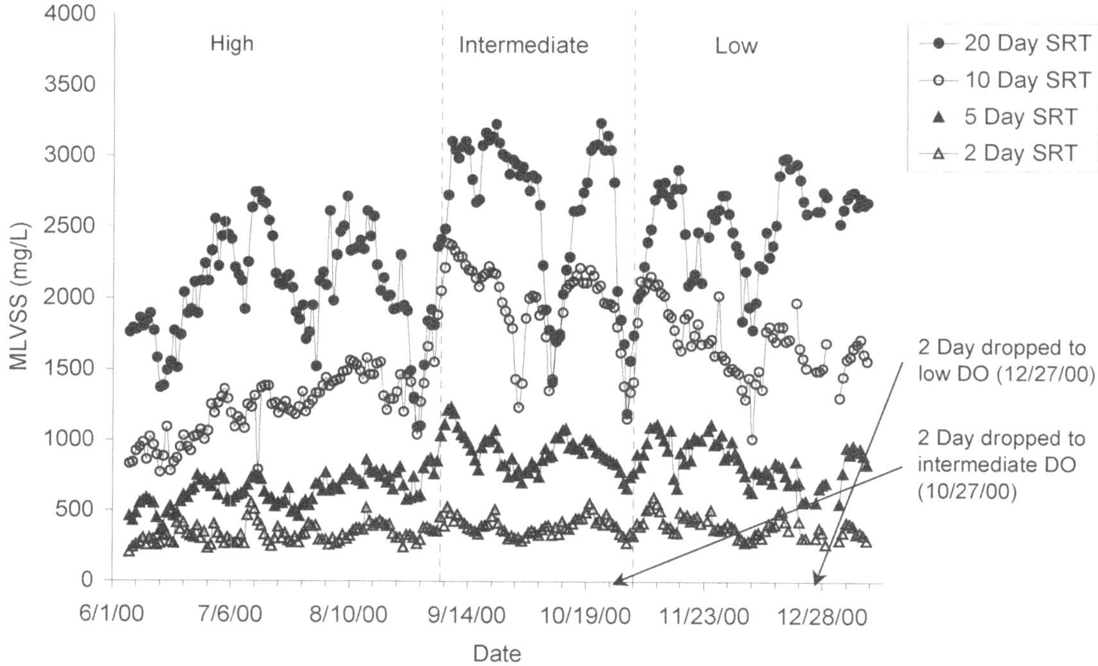

Figure 4-4. Fluctuations in MLVSS Concentration at Each SRT Due to Changes in DO and Influent COD Concentrations During 20°C Operation.

During the period from October 5th to November 2nd, solids concentrations, particularly in the 20-day and 10-day SRT reactors, underwent major changes that were attributed to a consistent foaming problem during this time. The foaming did not result in settling problems but often caused the reactor effluent port to clog. Solids would remain in the reactor until the foam at the overflow dispersed, and then a significant amount of liquid would flood into the clarifier. This overloading of the clarifier resulted in high effluent solids loss on several occasions, thereby affecting MLVSS concentrations.

Averages and standard deviations for the MLVSS concentrations during 20°C operation, excluding the two problem events, are shown in Table 4-7. In terms of maintaining a constant biomass concentration, it is clear from Table 4-7 that the shorter SRT reactors were more difficult to control at either high or low DO levels based on the larger relative standard deviations.

Table 4-7. Average (±SD) MLVSS Concentrations (mg/L) During Operation at 20°C.

DO Level	High	Intermediate	Low
20-Day SRT	2113 (±331)	2941 (±173)	2515 (±295)
10-Day SRT	1238 (±252)	2041 (±289)	1686 (±234)
5-Day SRT	644 (±73)	932 (±151)	854 (±149)
2-Day SRT	358 (±73)	407 (±83)	345 (±49)

Multiple linear regression analysis was performed on the MLVSS data, as discussed in Section 3.3.3, to determine how much of the variability in measured values could be attributed to changes in the controlled variables and which variables had the largest relative effect on solids levels. The developed model fit the experimental data quite well, as evidenced by a correlation coefficient of 0.926. This indicated that 85.8% of the observed variability in MLVSS concentrations could be explained by changes in the influent COD, SRT, DO, and the compound effects of DO and SRT.

Of the four independent variables included in the model, the null hypothesis (that the partial regression coefficient was zero) could only be rejected for the SRT at a 95% confidence interval. This result indicated that SRT was the prime determinant of the MLVSS concentration, and that variations in DO and influent COD levels did not significantly affect the MLVSS during 20°C operation. This result was anticipated since, as discussed in Section 2.3.1, for a fixed reactor volume, selection of the SRT determines the total biomass concentration in the system. Because the influent COD concentration remained fairly stable throughout the 20°C study (Figure 4-1), the influent COD level would not be expected to cause large variations in the MLVSS values. As noted in previous sections, changes in DO level were not likely to yield a significant impact on MLVSS concentrations, since the oxygen half-saturation constant for heterotrophs is quite low.

4.1.2.2 15°C Operation

In general, MLVSS concentrations were a function of SRT during 15°C operation of the municipal bench-scale reactors (Figure 4-5). However, there were some anomalies worth noting. The first major change in solids concentration occurred between May 3 and May 24, 2001, when an ink-like material was observed in the collection system of the full-scale municipal WWTP. A foaming problem also occurred between July 17 and July 26, 2001, resulting in a drastic drop in MLVSS concentrations with a concurrent rise in ESS for all reactors. Another anomaly occurred in the 5-day SRT reactor between September 2 and September 9, 2001. The cleaning arm in the external clarifier malfunctioned, causing the recycle lines to clog with solids. This problem was not discovered for 3 days, causing unusually low solids concentrations in the mixed liquor. Once the problem was corrected, however, the solids concentrations rapidly increased and returned to steady-state levels. With the exclusion of the previously described problem events, the MLVSS concentration in the 2-day SRT reactor remained fairly stable throughout the 15°C study. In the 20-, 10-, and 5-day SRT reactors, the MLVSS concentrations increased throughout operation at the high DO level, and then remained fairly stable throughout operation at the intermediate- and low-DO level periods. Averages and standard deviations for the MLVSS concentrations at 15°C (excluding the three anomalies) are shown in Table 4-8. The average MLVSS concentrations at 15°C are very similar to those reported at 20°C (Table 4-7), suggesting that the 5°C temperature decrease did not significantly affect MLVSS values.

Linear regression analysis of the 15°C MLVSS data yielded a model that described the experimental data very well (R=0.974) and revealed that 95% of the variability in MLVSS could be attributed to changes in the influent COD, SRT, DO and the interaction between SRT and DO. Again, the SRT was found to have the greatest impact on MLVSS. However, in this case, the interaction term describing the joint effect of SRT and DO also contributed significantly to the model. To study this interaction, the data was subdivided by DO level, and additional linear regression analyses were performed for each data subset. The resulting linear models showed

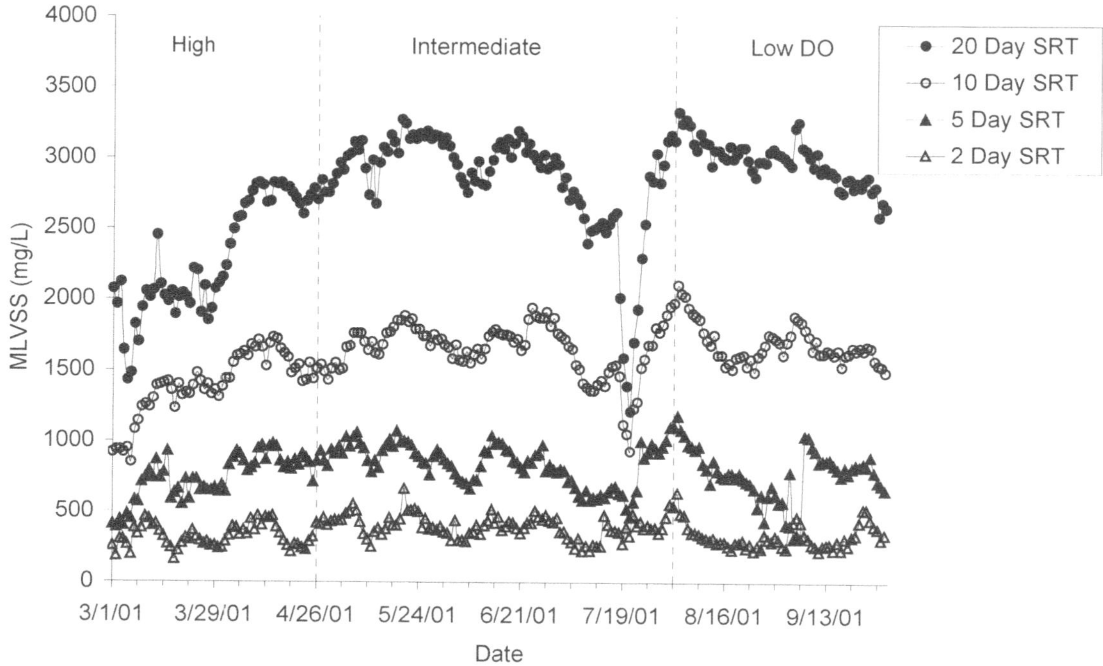

Figure 4-5. Fluctuations in MLVSS Concentration at Each SRT Due to Changes in DO and Influent COD Concentrations During 15°C Operation.

Table 4-8. Average (±SD) MLVSS Concentrations (mg/L) During Operation at 15°C.

DO Level	High	Intermediate	Low
20-Day SRT	2293 (±398)	2838 (±380)	2974 (±156)
10-Day SRT	1398 (±223)	1645 (±188)	1691 (±141)
5-Day SRT	739 (±158)	838 (±143)	772 (±186)
2-Day SRT	321 (±79)	395 (±80)	330 (±90)

that the impact of SRT on the MLVSS increased as the DO level decreased. The partial regression coefficients indicated that an increase in the SRT of 1 day resulted in an increase in MLVSS concentration of 110 mg/L, 125 mg/L, and 139 mg/L at the high, intermediate, and low DO levels, respectively. Table 4-8 confirms that throughout the study, the difference in MLVSS concentrations between each SRT increased. The influent COD concentration was not found to significantly affect the MLVSS concentration, which is not surprising since the strength of the influent wastewater did not vary substantially throughout the 15°C study period (Figure 4-2).

4.1.2.3 10°C Operation

During 10°C operation of the municipal bench-scale reactors, the MLVSS levels appeared to mirror the trend in influent COD concentration as indicated by a comparison of Figures 4-3 and 4-6. The influent COD concentration decreased throughout the 10°C operational period and the MLVSS values decreased in a corresponding fashion. Although Table 4-9 appears to indicate that lower DO levels decreased MLVSS concentrations, it is believed that the lower influent COD levels were responsible for these results.

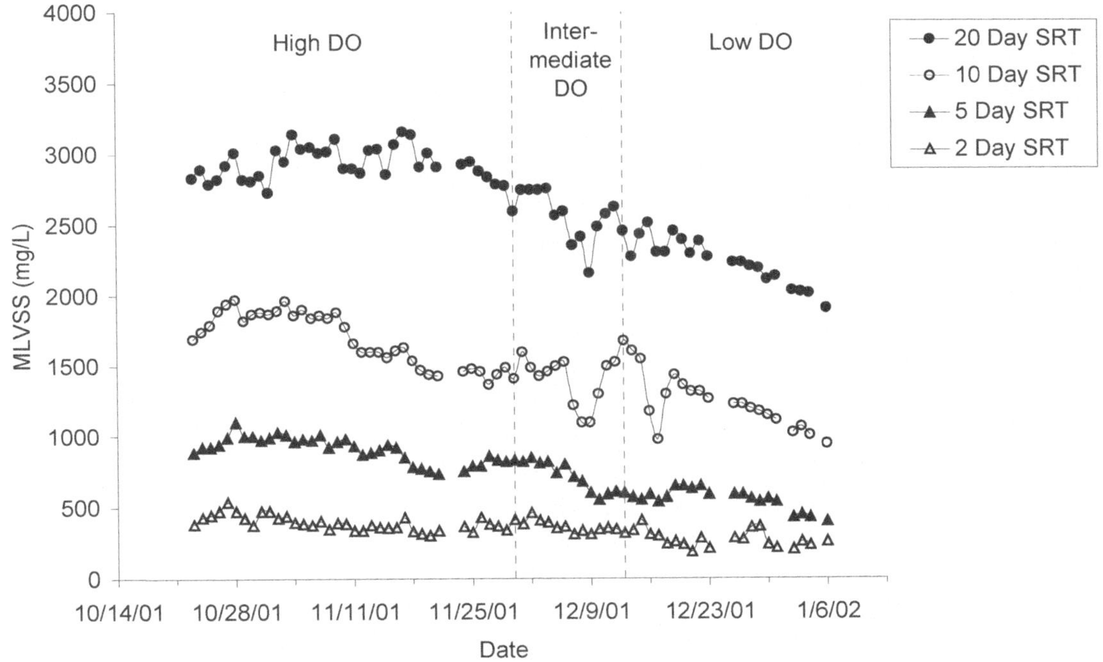

Figure 4-6. Fluctuations in MLVSS Concentration at Each SRT Due to Changes in DO and Influent COD Concentrations During 10°C Operation.

Table 4-9. Average (±SD) MLVSS Concentrations (mg/L) During Operation at 10°C.

DO Level	High	Intermediate	Low
20-Day SRT	2939 (±114)	2571 (±178)	2252 (±230)
10-Day SRT	1698 (±187)	1398 (±166)	1247 (±201)
5-Day SRT	908 (±92)	724 (±107)	556 (±73)
2-Day SRT	391 (±55)	368 (±43)	276 (±58)

The average influent COD concentration measured during operation at the high DO level was consistent with values measured throughout the 15°C and 20°C studies. The MLVSS levels during these periods were also similar. This suggests that the decrease in temperature did not affect the MLVSS concentration.

The multiple linear regression model developed for the 10°C MLVSS data described the experimental results very well (R=0.99), and revealed that 98.3% of the observed variability in MLVSS values could be attributed to changes in controlled variables. Only the SRT and the compound effects of SRT and DO were found to affect the MLVSS in a statistically significant manner. It was expected that the influent COD level would be a significant predictor of MLVSS concentration. However, it is believed that this was not observed in the model due to problems of multicollinearity. Since the influent COD and DO decreased in a corresponding fashion, these parameters were highly correlated. The inclusion of both terms over-specified the model, making the effect of the individual variables on the independent variable unclear. To test this theory, the data was subdivided based on DO, and additional analyses were performed. The results of these models showed that the influent COD was not significant at the high DO level, but did impact the MLVSS at the intermediate and low DO levels. This is reasonable since at the high DO level, the influent COD concentration was much more stable than at the other DO levels (Figure 4-3). At the high DO level, only the SRT controlled the MLVSS concentration. Even though the influent COD concentration impacted the MLVSS at the intermediate and low DO levels, SRT exerted much more influence on the solids concentration under these conditions.

As shown in equation 2.3-6, the MLVSS concentration is directly proportional to the amount of COD removed in the treatment system. Throughout the 10°C study, the decrease in influent COD concentration resulted in a decrease in the amount of COD removed. In turn, a decrease in MLVSS values was observed. Several other studies have shown a correlation between influent COD level and MLVSS concentration. Jianlong et al. (2000) conducted a study on the effect of organic loading rates on treatment performance in a hybrid biological reactor treating domestic wastewater that was supplemented with glucose to control the loading rates. In this hybrid biological reactor system, the activated sludge process was combined with a fixed film biomass process. Average influent COD levels of 190 mg/L, 365 mg/L, and 760 mg/L resulted in average MLVSS values in the suspended biomass of 1900 mg/L, 2050 mg/L, and 2250 mg/L, respectively, and MLVSS values in the total biomass of 4300 mg/L, 4850 mg/L, and 5750 mg/L, respectively. Lee et al. (2000) also performed experiments to investigate the effect of organic loading on treatment performance in hybrid biological reactors and obtained similar results. In their study, a bench-scale conventional activated sludge reactor was used as a control. When the reactor was operated at an SRT of 10 days and used to treat synthetic wastewater with COD levels of 250 mg/L, 500 mg/L, and 1000 mg/L, MLVSS concentrations of 2080 mg/L, 2120 mg/L, and 2200 mg/L were measured. In these studies, and the current investigation, the MLVSS concentration was positively correlated with influent COD levels.

4.1.2.4. Impact of Test Parameters on Solids Variability

Multiple linear regression analysis was performed on all MLVSS data from the 20°C, 15°C, and 10°C operating periods. Results showed that 92% of the observed variability in MLVSS values could be explained by the influent COD, SRT, temperature, DO, and the two-way and three-way interactions of these independent variables. The partial correlation coefficients showed that the independent variable that had the greatest impact on the MLVSS was the SRT. This result was expected since more organisms are retained in the system at longer SRTs thereby resulting in larger MLVSS values. The individual effects of temperature, DO, and influent COD were not found to be statistically significant when all the data were considered.

These findings suggest that SRT controls the MLVSS concentration as long as the influent COD level is fairly consistent. However, large variations in influent COD concentrations are reflected in the MLVSS concentration. DO and temperature were not found to significantly impact the MLVSS concentration. These findings are consistent with reported results found in the literature. Chuang et al. (1997) observed that MLSS concentrations in a combined activated sludge-biofilm system were not affected by DO concentration. When operated at a 5-day SRT, MLSS concentrations in the system were 920 mg/L, 1010 mg/L, and 1030 mg/L, at DO levels of 2.0 mg/L, 0.5 mg/L, and 0.1 mg/L, respectively. Lau et al. (1984) reported similar findings for an activated sludge CSTR operated at an SRT of 0.33 days. At DO levels of 7.04 mg/L and 0.44 mg/L, average MLSS concentrations of 519 mg/L and 491 mg/L, respectively were measured in the system. McClintock et al. (1993) measured MLVSS concentrations of 1284 mg/L, 1050 mg/L, and 1373 mg/L at temperatures of 20°C, 15°C, and 10°C, respectively, in a pilot activated sludge system treating domestic wastewater operated at an SRT of 5 days, indicating that temperature did not significantly affect MLVSS concentration.

4.1.3 Sludge Settling and Effluent Suspended Solids

4.1.3.1 20°C Operation

SVI values for each reactor during the 20°C treatment study are displayed in Figure 4-7. Sludge settling was excellent throughout most of the treatment study. Average SVI values for each reactor, operating at steady state, are shown in Table 4-10. Although DO can have a negative impact on settling, SRT and substrate utilization appear to be the main factors which affect sludge settling (Surucu and Cetin, 1990). DO contributes to poor SVI values when sufficient organic loading (and subsequent substrate utilization) occurs. Palm et al. (1980) demonstrated that an organic removal rate of 0.2 mg to 0.3 mg COD/mg MLVSS/day at DO concentrations of 0.1 mg/L to 0.5 mg/L did not negatively affect SVI. However, organic removal rates above this range did result in significant settling problems. For DO concentrations between 0.5 mg/L and 1.0 mg/L, a substrate removal rate greater than 0.45 mg COD/mg MLVSS/day was required to affect SVI values. The average substrate removal rate for the 20-day SRT reactor in this study, operating at 0.5 mg/L DO concentration, was calculated to be 0.10 mg COD/mg MLVSS/day. This value is well below the organic removal rate that leads to poor sludge settling.

Figure 4-7. Effect of DO Concentration and SRT on Settling Performance During 20°C Operation.

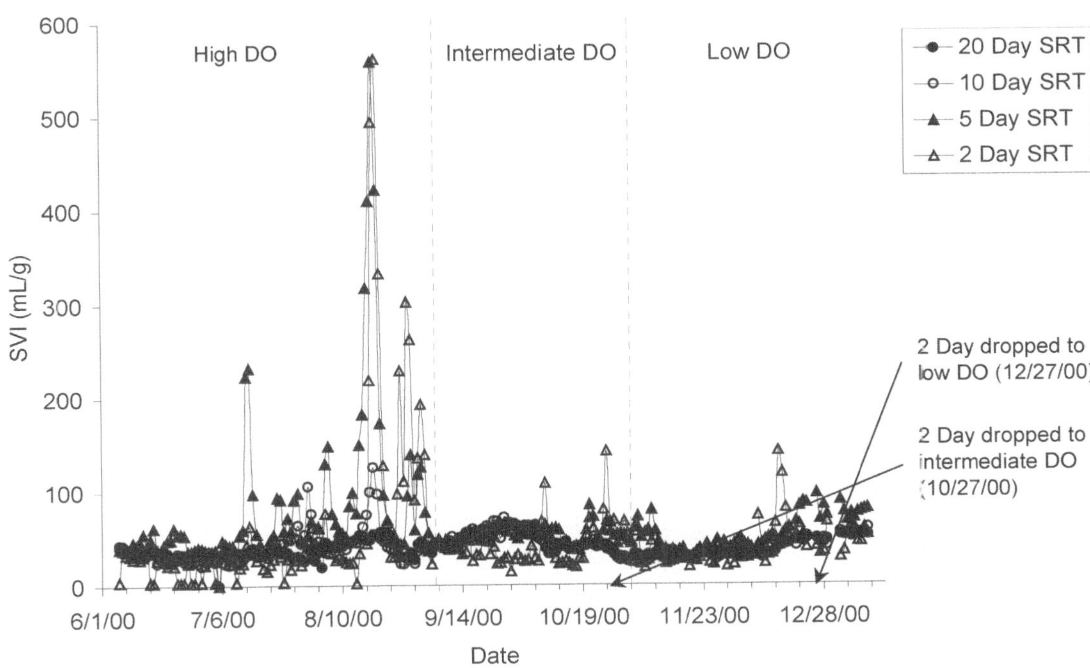

Table 4-10. Average (±SD) SVI Values (mL/g) During 20°C Operation.

DO Level	High	Intermediate	Low
20-Day SRT	37 (±8)	47 (±12)	37 (±9)
10-Day SRT	39 (±19)	48 (±12)	37 (±10)
5-Day SRT	81 (±86)	52 (±9)	55 (±19)
2-Day SRT	53 (±77)	42 (±24)	49 (±12)

For the 10-day SRT reactor, the average substrate utilization rates for the 0.5 mg/L and 0.2 mg/L DO operational periods were both 0.12 mg COD/mg MLVSS/day, well below the organic removal rate necessary to cause settling problems. The average substrate removal rates in the 5-day SRT reactor were slightly higher during the 0.5 mg/L and 0.2 mg/L DO periods of operation (0.23 mg and 0.26 mg COD/mg MLVSS/day, respectively). Once again, these levels were well below the boundary for sludge bulking. The average substrate removal rate in the 2-day SRT reactor was 0.57 mg COD/mg MLVSS/d, a value that exceeds the upper substrate removal rate boundary for effective settling at a DO concentration of 0.5 mg/L. However, settling problems did not occur at this SRT and DO level. It is possible that the short SRT limited growth of the filamentous organisms caused settling problems (Wanner, 1998).

Figure 4-8 depicts ESS over the time course of the 20°C study. Average ESS values for each of the four reactors were below 30 mg/L, the average monthly discharge limit for the full-scale municipal WWTP. However, there were a couple of periods in which ESS levels were high. The first major incident occurred when the influent

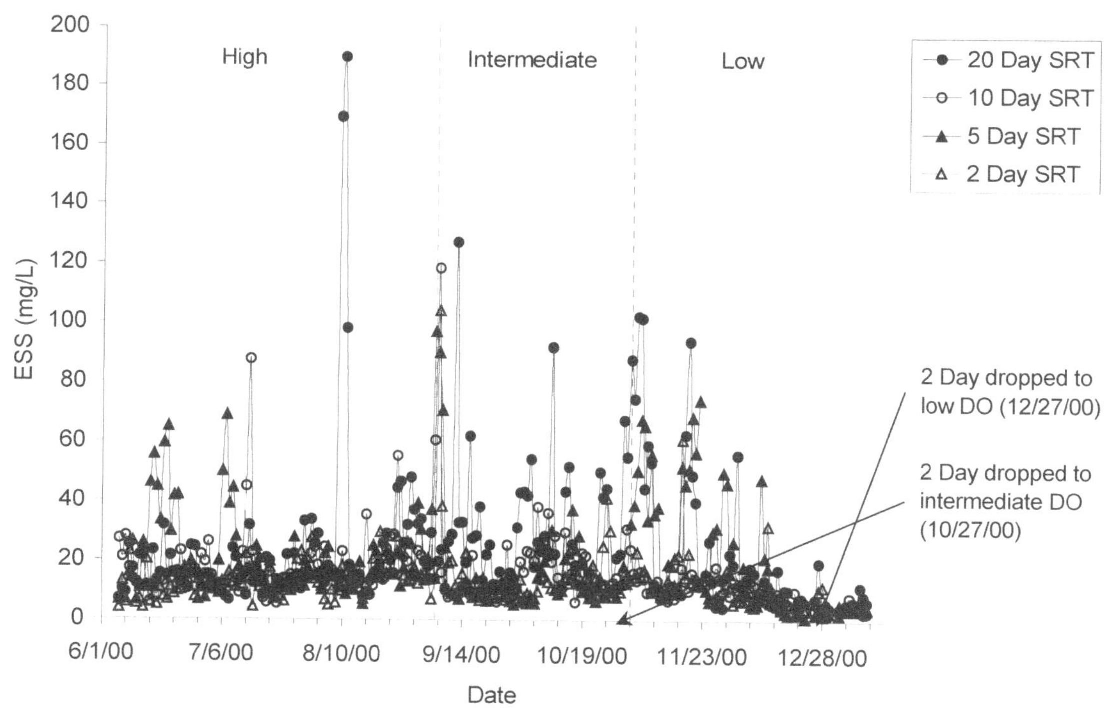

Figure 4-8. Effect of DO Concentration and SRT on Effluent Suspended Solids During 20°C Operation.

pump tubing was replaced with thick-walled tubing. The high ESS levels measured were consistent with the MLVSS drop in the four reactors (particularly the 20- and 10-day SRT reactors). The second period of poor settling occurred from November 1 to November 21, 2001, primarily in the 20- and 5-day SRT reactors. Once again, high ESS values corresponded to drastically lower MLVSS levels. As stated earlier, foaming during significant portions of October and into November likely contributed to excessive ESS discharges. Average ESS values for the three operational periods (high, intermediate, and low DO) are reported in Table 4-11.

Table 4-11. Average (±SD) ESS Values (mg/L) During 20°C Operation.

DO Level	High	Intermediate	Low
20-Day SRT	23 (±26)	30 (±23)	20 (±24)
10-Day SRT	18 (±12)	17 (±16)	9 (±4)
5-Day SRT	21 (±16)	15 (±15)	20 (±21)
2-Day SRT	14 (±10)	12 (±9)	11 (±9)

4.1.3.2 15°C Operation

SVI data for the 15°C temperature are shown in Figure 4-9 and tabulated in Table 4-12 for each reactor and DO concentration. The 20-day SRT reactor SVI remained fairly constant during 15°C operation. The 10-day SRT reactor SVI values increased steadily throughout the three 15°C DO phases, approximately doubling when the DO concentration was dropped to 0.2 mg/L. The 5- and 2-day SRT reactors had much higher and more variable SVI values than the 10- and 20-day reactors.

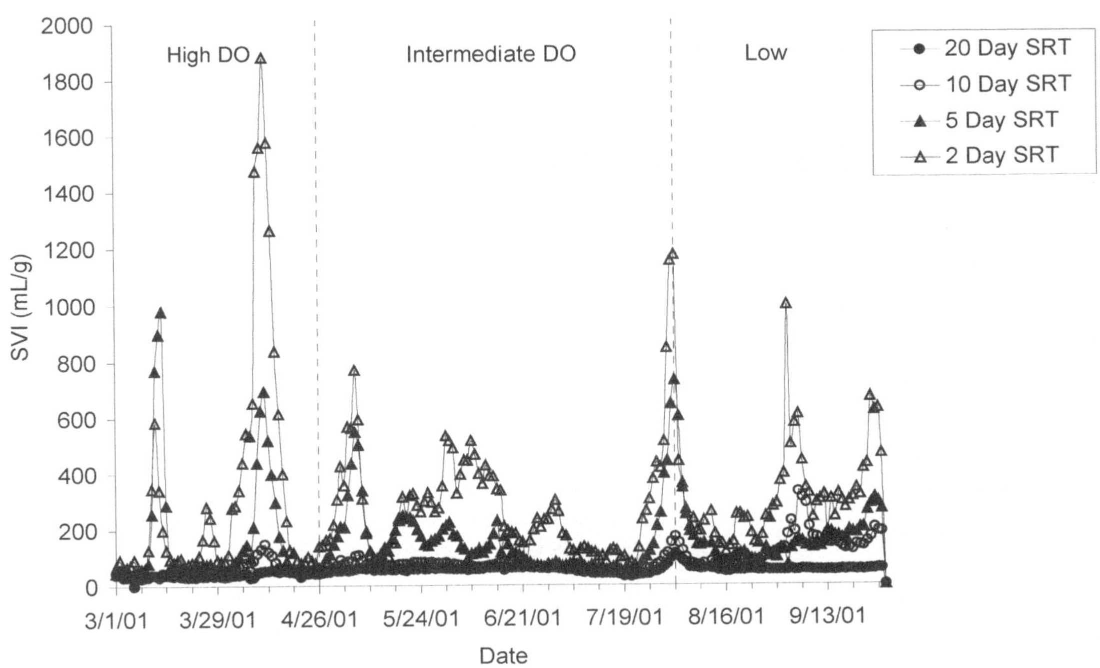

Figure 4-9. Effect of DO Concentration and SRT on Settling Performance During 15°C Operation.

Table 4-12. Average (±SD) SVI Values (mL/g) During 15°C Operation.

DO Level	High	Intermediate	Low
20-Day SRT	40 (±7)	55 (±10)	56 (±14)
10-Day SRT	54 (±26)	68 (±19)	132 (±61)
5-Day SRT	168 (±225)	151 (±111)	173 (±116)
2-Day SRT	322 (±432)	275 (±181)	331 (±200)

Figure 4-10 depicts ESS over the 15°C study. Average ESS values for each of the four reactors were below 30 mg/L (Table 4-13), the average monthly discharge limit for the full-scale municipal WWTP.

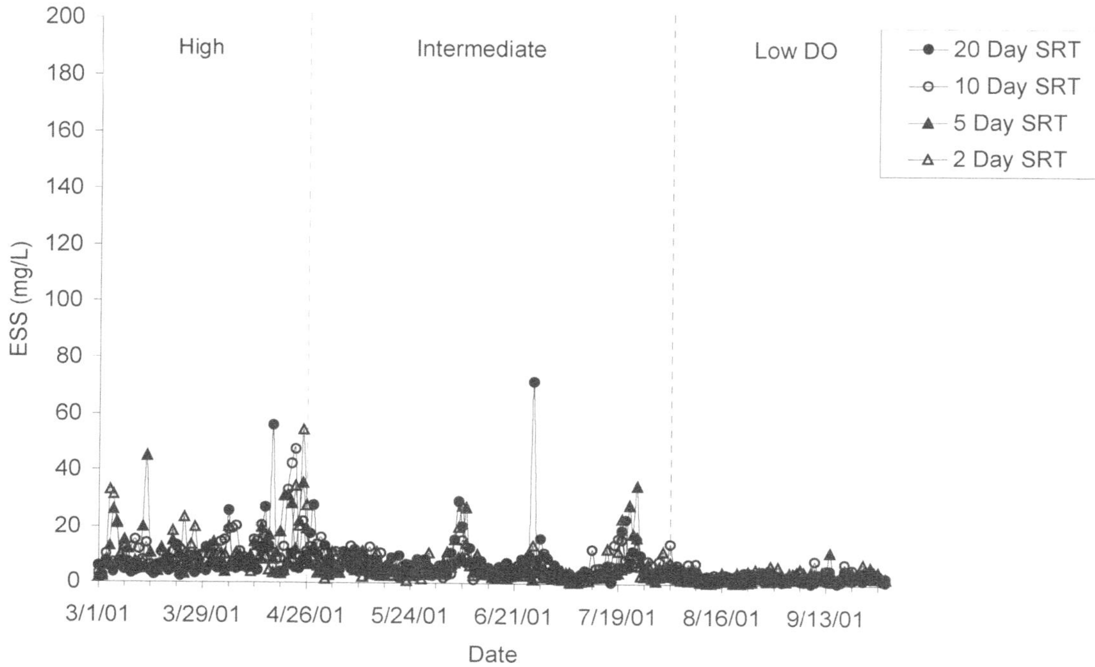

Figure 4-10. Effect of DO Concentration and SRT on Effluent Suspended Solids During 15°C Operation.

Table 4-13. Average (±SD) ESS Values (mg/L) During 15°C Operation.

DO Level	High	Intermediate	Low
20-Day SRT	7 (±8)	8 (±8)	3 (±1)
10-Day SRT	12 (±8)	6 (±4)	3 (±2)
5-Day SRT	12 (±9)	6 (±5)	2 (±2)
2-Day SRT	11 (±9)	6 (±5)	3 (±1)

4.1.3.3 10°C Operation

During 10°C operation, SVI values for the shorter SRT systems were much higher than those of the longer SRT systems (Figure 4-11). In general, SVI values for all SRTs increased as temperature was decreased to 10°C. Table 4-14 provides average SVI values for each reactor during this period. The 5-day reactor was particularly affected by the drop in temperature to 10°C. SVI has been found to be a function of SRT and COD loading rate (Surucu and Cetin, 1990). This was also true in this study. Settling efficiency decreased with decreasing SRT and increasing COD loading. The COD loading to the 2- and 5-day SRT reactors was higher than that of the 10- and 20-day SRT reactors due to the lower MLVSS concentrations maintained in the short SRT reactors. Temperature also seems to be a major factor in settling as the lower temperature yielded higher SVI values.

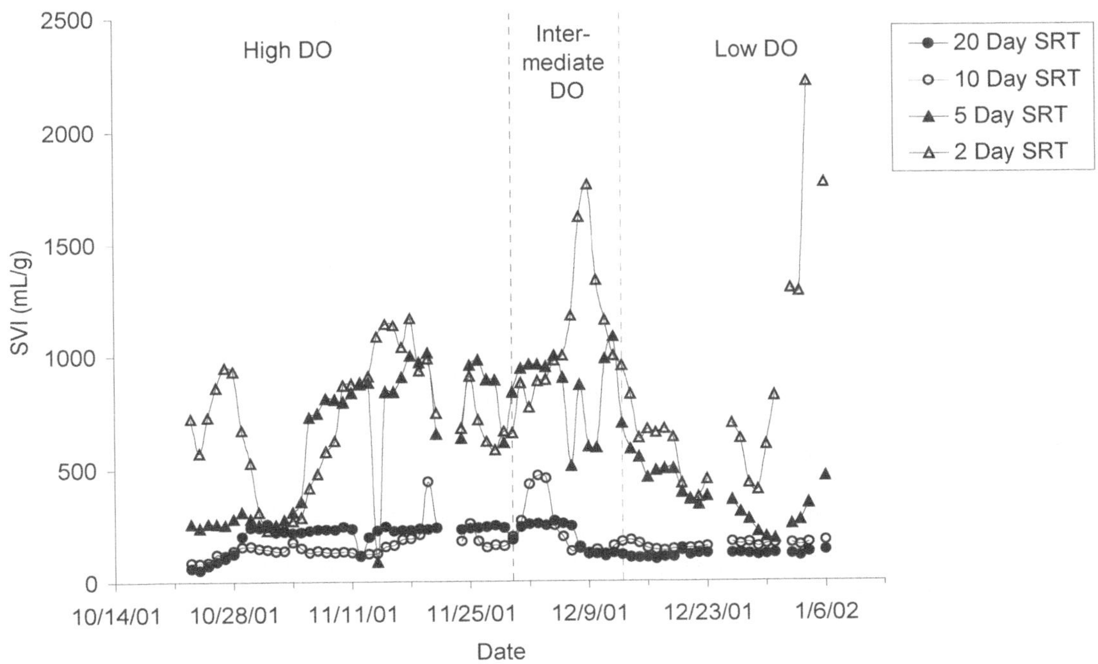

Figure 4-11. Effect of DO Concentration and SRT on Settling Performance During 10°C Operation.

Table 4-14. Average (±SD) SVI Values (mL/g) During 10°C Operation.

DO Level	High	Intermediate	Low
20-Day SRT	205 (±60)	201 (±63)	118 (±56)
10-Day SRT	158 (±62)	238 (±132)	161 (±89)
5-Day SRT	601 (±305)	861 (±179)	390 (±277)
2-Day SRT	712 (±276)	1088 (±323)	806 (±442)

ESS data are shown in Figure 4-12 and average values are presented in Table 4-15 for each reactor at each 10°C operational phase. The ESS were not affected by temperature or DO concentration. Averages were all below the monthly discharge limit for full-scale municipal WWTPs (30 mg/L).

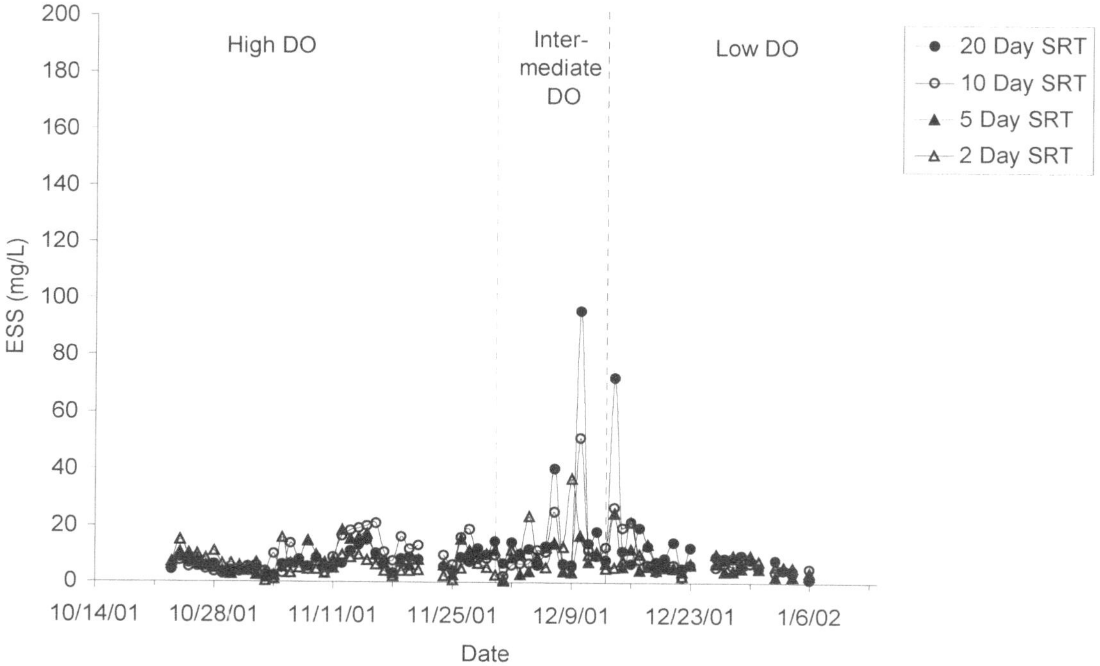

Figure 4-12. Effect of DO Concentration and SRT on Effluent Suspended Solids During 10°C Operation.

Table 4-15. Average (±SD) ESS Values (mg/L) During 10°C Operation.

DO Level	High	Intermediate	Low
20-Day SRT	7 (±3)	20 (±24)	12 (±19)
10-Day SRT	9 (±6)	12 (±13)	8 (±6)
5-Day SRT	8 (±4)	8 (±6)	6 (±5)
2-Day SRT	6 (±4)	6 (±5)	8 (±4)

4.1.4 Nitrification Performance

Influent ammonia and reactor effluent ammonia, nitrite, and nitrate concentrations (as N) were used to calculate ammonia oxidation rates and efficiencies. Nitrite- and nitrate-N are products of the nitrifying bacteria, while ammonia-N in the effluent represents nitrogen not biologically utilized. Thus, the ammonia oxidation rate for the AOB can be calculated using these effluent inorganic nitrogen concentrations as shown in equation 4.1.4-1.

$$\text{NH}_4^+ \text{ Oxidation Rate} = \frac{Q\left(\text{Eff. NO}_2^- + \text{Eff. NO}_3^-\right)}{V} \qquad (4.1.4\text{-}1)$$

where:

NH_4^+ Oxidation Rate = rate of ammonia oxidation (mg N/L/day)
Q = volumetric flow rate (L/day) = 27.4 L/day
Eff. NO_2^- = effluent nitrite (mg/L as N)
Eff. NO_3^- = effluent nitrate (mg/L as N)
V = reactor volume = 10 L

The conversion efficiency for the AOB can be calculated using these effluent inorganic nitrogen concentrations as shown in equation 4.1.4-2.

$$\text{NH}_4^+ \text{ Oxidation Eff.}(\%) = \frac{\left(\text{Eff. NO}_2^- + \text{Eff. NO}_3^-\right)}{\left(\text{Eff. NH}_4^+ + \text{Eff. NO}_2^- + \text{Eff. NO}_3^-\right)} \times 100\% \qquad (4.1.4\text{-}2)$$

where:

Eff. NO_2^- = effluent nitrite (mg/L as N)
Eff. NO_3^- = effluent nitrate (mg/L as N)
Eff. NH_4^+ = effluent ammonia (mg/L as N)

Nitrite oxidation rates and efficiencies can be computed in a similar manner and will be discussed in Section 4.1.4.3.

4.1.4.1 Ammonia Removal

4.1.4.1.1 20°C Operation

Figures 4-13(A) through 4-16(A) show the influent ammonia and reactor effluent ammonia, nitrite, and nitrate concentrations (as N) for the 20°C study as a function of DO level over time for the 20-, 10-, 5-, and 2-day SRT systems, respectively. Average ammonia concentrations and corresponding ammonia oxidation rates and efficiencies for each DO level in each SRT reactor are presented in Table 4-16 for the 20°C study. The rates shown in this table are comparable to the nitrification rates measured by Ydstebo et al. (2000), which ranged from 42.6 mg to 53.6 mg N/L/day in a Norwegian BNR treatment facility treating domestic wastewater with an average influent total nitrogen concentration of 32 mg/L. Prinčič et al. (1998) reported ammonia oxidation rates of 37 mg to 74 mg N/L/day for enriched nitrifying bacterial cultures in continuously fed reactors operated under excess oxygen conditions. McClintock et al. (1993) reported similar results in a pilot-scale activated sludge system treating domestic wastewater with influent TKN concentrations of 25 mg to 37 mg N/L operated at SRTs between 1.5 and 15 days. In their study, ammonia oxidation rates of 45 mg to 81 mg N/L/day were measured in the reactors and corresponded to nitrification efficiencies between 65% and 100%.

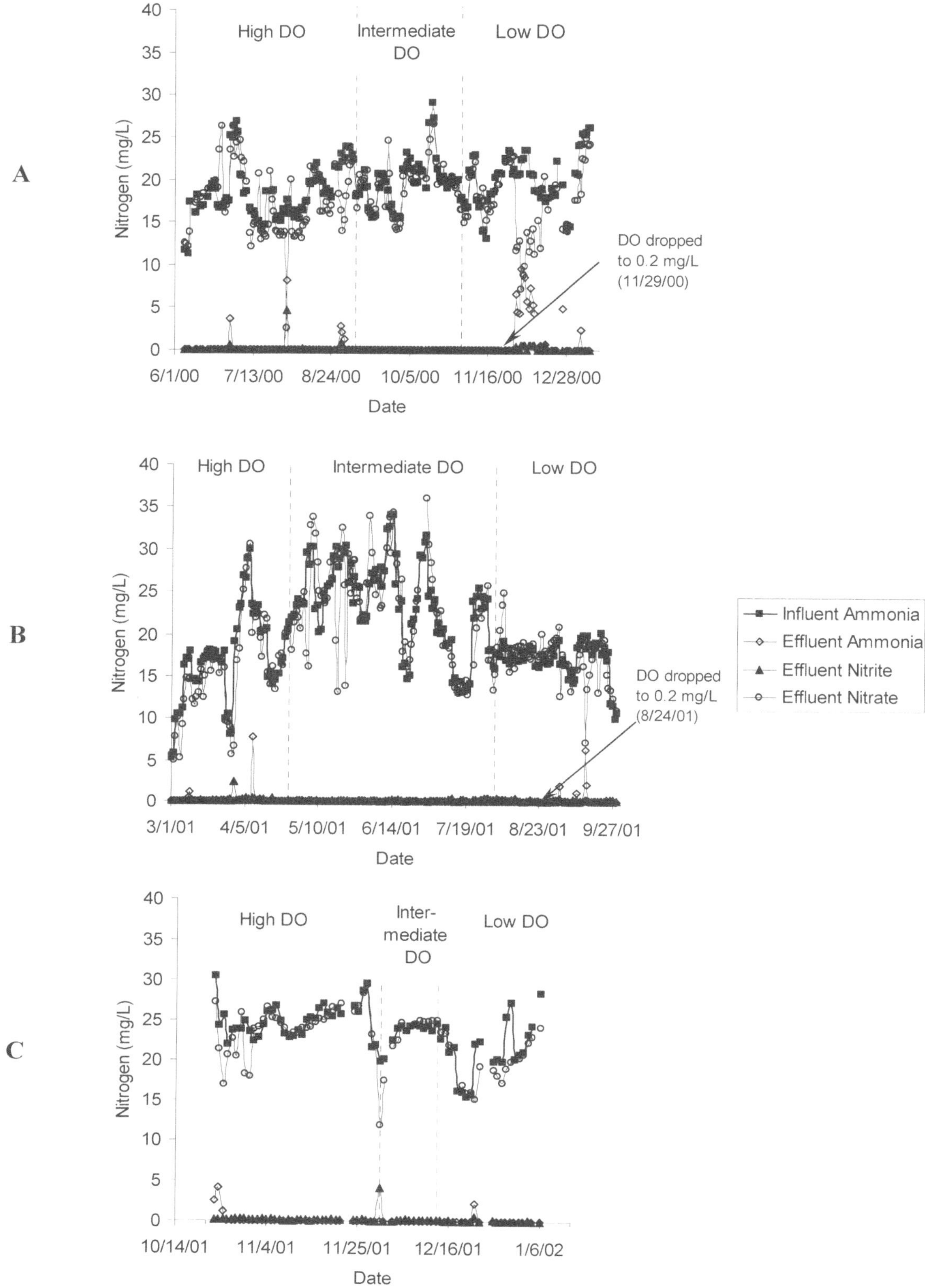

Figure 4-13. 20-day SRT Reactor Nitrogen Results (A) 20°C, (B) 15°C, (C) 10°C.

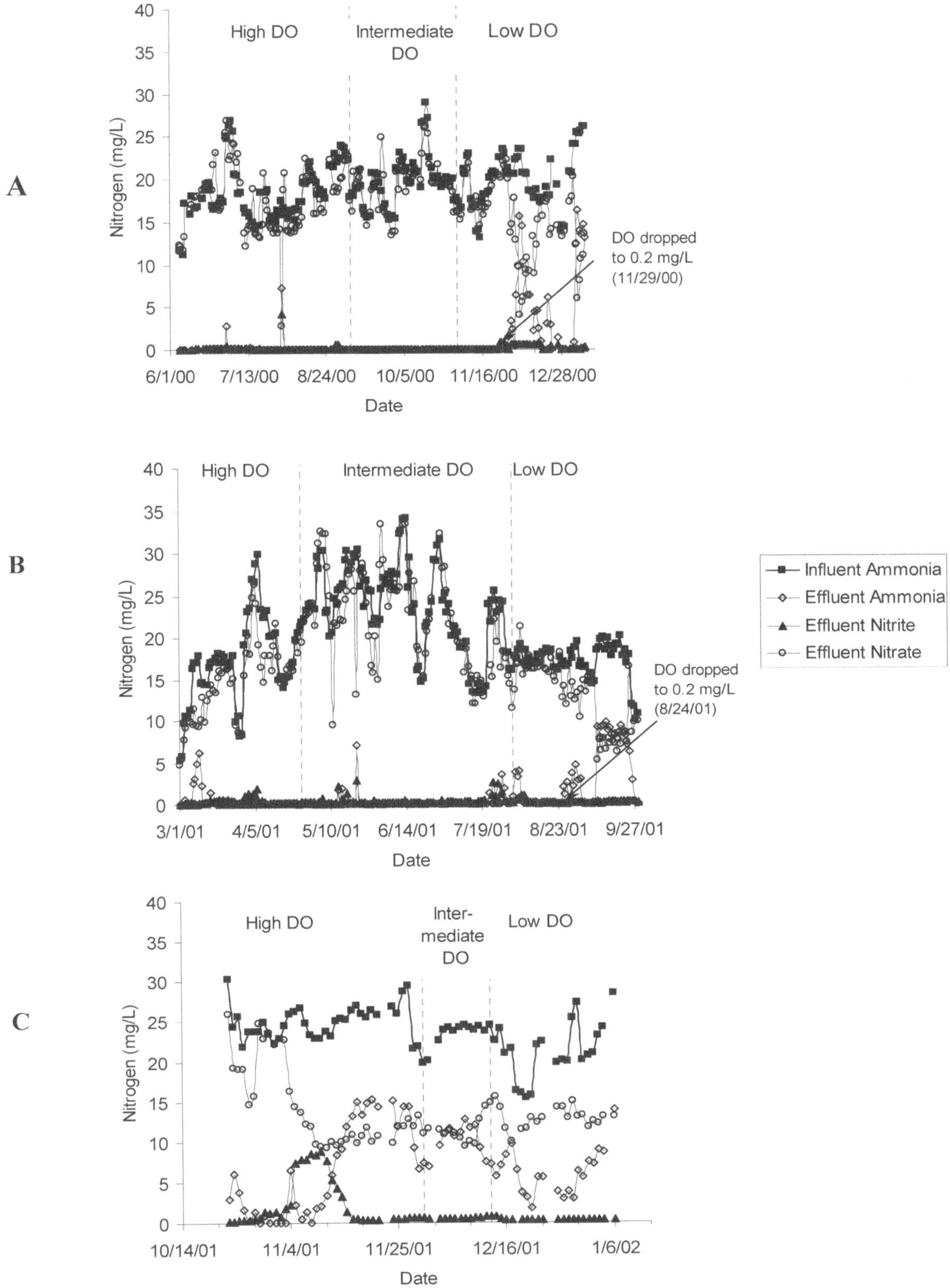

Figure 4-14. 10-day SRT Reactor Nitrogen Results (A) 20°C, (B) 15°C, (C) 10°C.

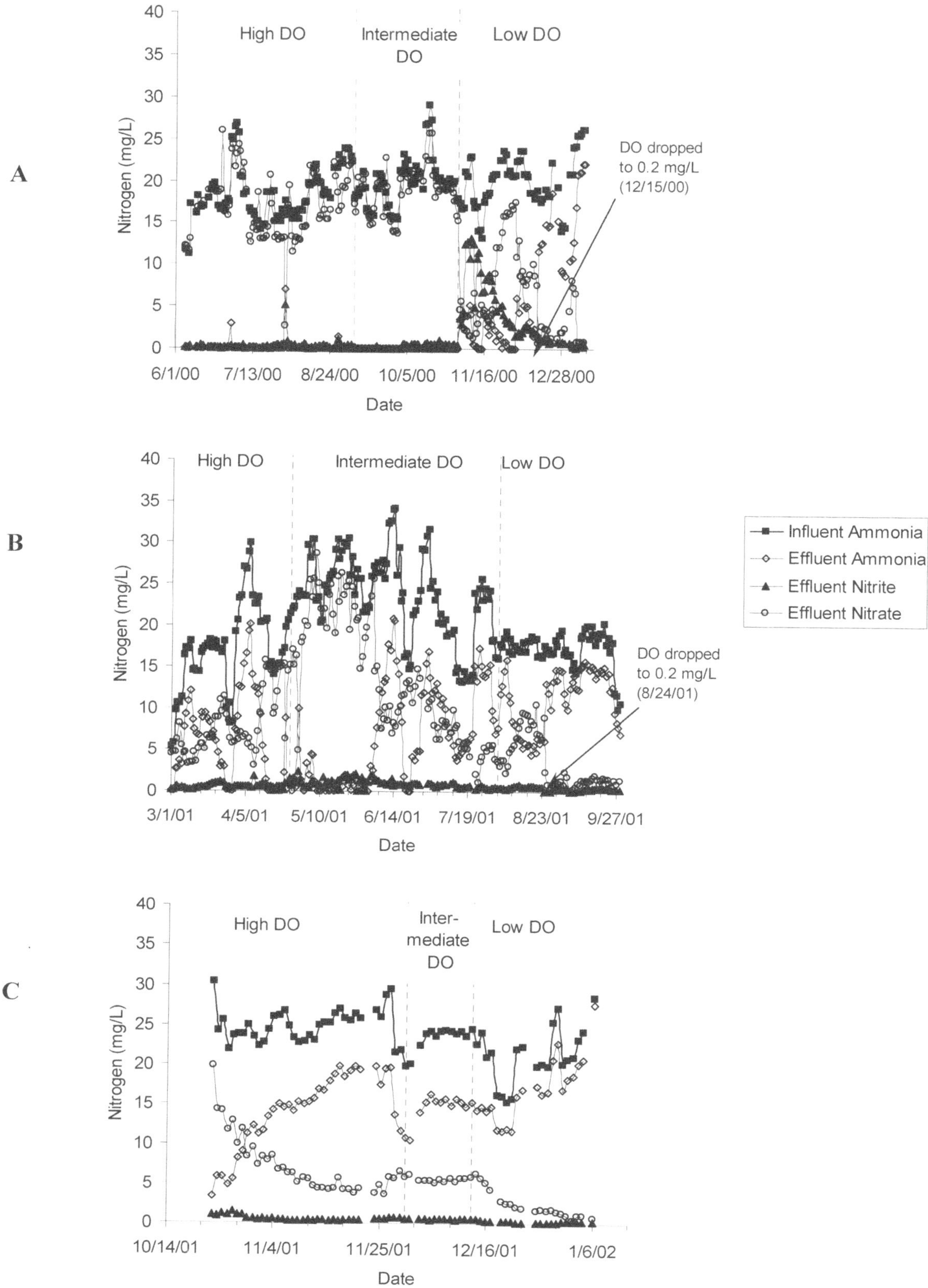

Figure 4-15. 5-day SRT Reactor Nitrogen Results (A) 20°C, (B) 15°C, (C) 10°C.

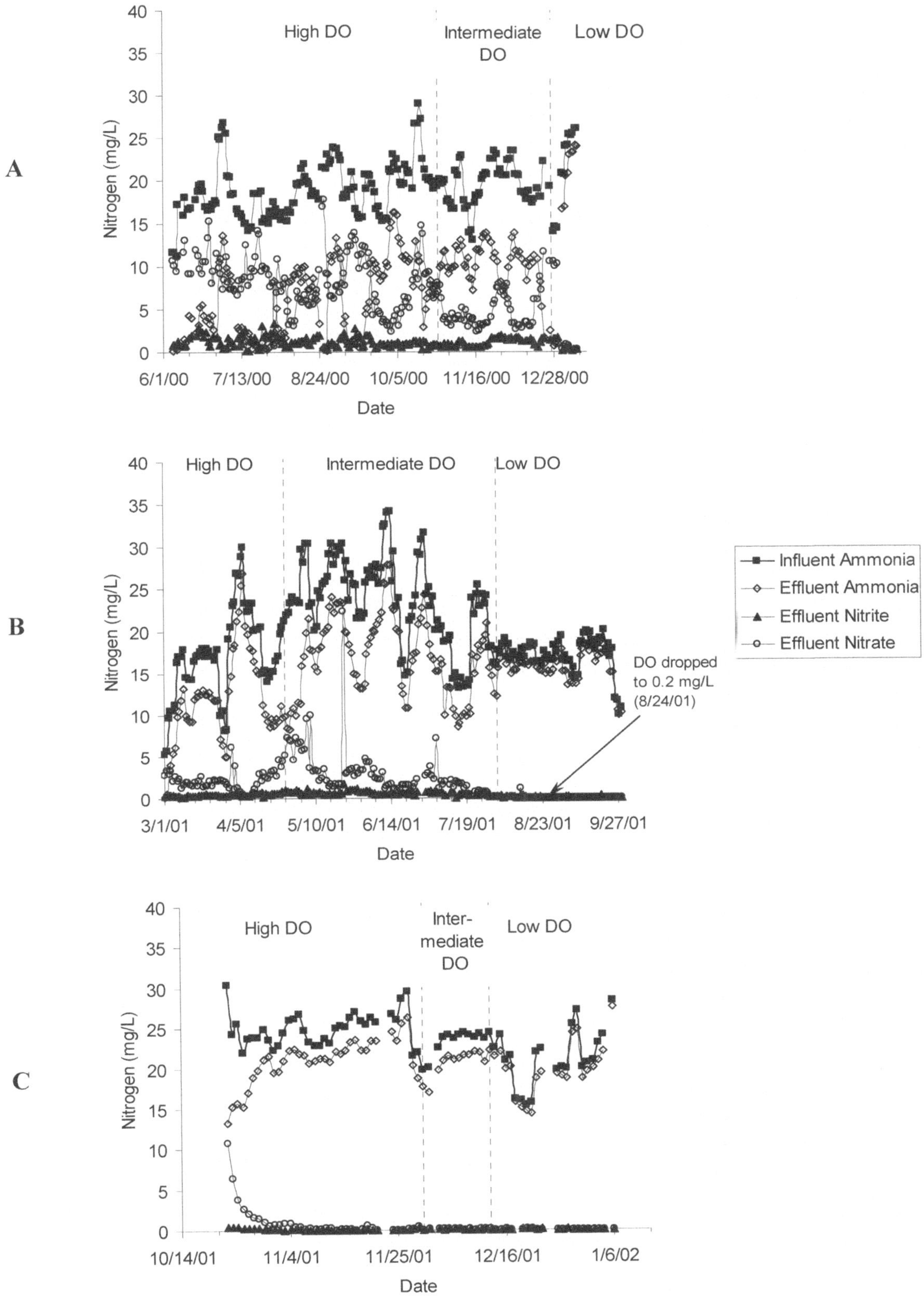

Figure 4-16. 2-day SRT Reactor Nitrogen Results (A) 20°C, (B) 15°C, (C) 10°C.

High rates and efficiencies of ammonia oxidation were consistently achieved in the 20-, 10-, and 5-day SRT reactors during operation at the high and intermediate DO levels (Table 4-16). The multiple linear regression model developed using data obtained under these operating conditions showed that only the influent ammonia significantly affected the ammonia oxidation rate at a 95% confidence level, indicating that nitrification performance was not substantially improved by increasing the SRT above 5 days or the DO concentration above 1 mg/L at a temperature of 20°C.

Ammonia removal did not decline in the 5-day SRT reactor until the DO level was lowered to 0.5 mg/L on November 2, 2000, and in the 20- and 10-day SRT reactors until the DO concentration was decreased to 0.2 mg/L on November 29, 2000. Prior to these changes in operating conditions, there were only a few occasions when ammonia was present in the effluent of these reactors. The first resulted from a spike in the ammonia concentration (from 17.4 mg/L to 25.1 mg/L as N) on July 1, 2000, which temporarily affected ammonia removal in all reactors. A second incident occurred on July 31, 2000, when a surfactant material passed through the system and also disrupted COD removal efficiency as previously discussed. Ammonia was also discharged from August 29 to August 31, 2000, when the installation of improper tubing resulted in decreased solids concentration. Data collected during these three perturbation events were excluded from all subsequent analyses.

Once the reactor DO concentration was dropped to 0.2 mg/L, complete ammonia removal was no longer consistently achieved at SRTs of 20 and 10 days, as evidenced by the routine presence of ammonia in the effluent of these reactors (Figures 4-13A, 4-14A). A considerable decrease in average ammonia oxidation rate was observed at an SRT of 5 days during the low DO period (November 3, 2000, to January 10, 2001). When operated at a DO concentration of 0.2 mg/L (January 6 to January 10, 2001) near complete nitrification failure occurred in this reactor (Figure 4-15A). Multiple linear regression analysis was performed on data collected from the 20-, 10-, and 5-day SRT reactors operating at each DO level. Results indicate that both the DO level and influent ammonia concentration significantly impacted the ammonia oxidation rate at a 95% confidence interval. Since the DO level did not significantly impact ammonia oxidation rates at these SRTs when only the high and intermediate DO levels were considered, this result suggests that DO levels ≤0.5 mg/L negatively impacted ammonia oxidation rate at 20°C.

Nitrification performance at an SRT of 2 days was significantly poorer than at the higher SRTs, as evidenced by the much lower ammonia oxidation rates in this reactor throughout the 20°C study (Table 4-16). Ammonia oxidation in the 2-day SRT reactor was also found to decrease with decreasing DO levels (Table 4-16). Multiple linear regression analysis of only the 2-day SRT reactor data showed that DO level significantly influenced the ammonia oxidation rate at a 95% confidence interval.

Near complete ammonia removal appears to have occurred at an SRT of 2 days from December 23, 2000, to January 1, 2001 (Figure 4-16A). However these results were attributed to a malfunction in the DO control system. This time period was excluded from the calculations of ammonia oxidation rate. During a short period, this reactor was operated at a DO of 0.5 mg/L (December 28, 2000, to January 10, 2001). With the exception of the above mentioned anomaly, complete nitrification failure occurred (Figure 4-16A).

During operation at the low DO level, SRT was found to have a statistically significant impact on ammonia oxidation rate at a 95% confidence interval based on multiple linear regression analysis. As expected, the impact of DO was found to decrease as SRT increased.

Table 4-16. Average (±SD) Ammonia Removal During Operation at 20°C.

DO Level	Parameter	20 day	10 day	5 day	2 day	Influent
High	Avg. NH_4^+ (mg N /L)	0.2 (±1.0)	0.1 (±0.8)	0.1 (±0.8)	6 (±4)	18 (±3)
	NH_4^+ Oxidation Rate (mg N/L/day)	46 (±13)	47 (±13)	46 (±13)	26 (±10)	n/a
	NH_4^+ Oxidation Efficiency (%)	99 (±6)%	99 (±6)%	99 (±5)%	64 (±24)%	n/a
Intermediate	Avg. NH_4^+ (mg N /L)	<0.1	<0.1	<0.1	11 (±2)	20 (±3)
	NH_4^+ Oxidation Rate (mg N/L/day)	52 (±10)	52 (±11)	51 (±10)	14 (±6)	n/a
	NH_4^+ Oxidation Efficiency (%)	100 (±0)%	100 (±0)%	100 (±0)%	34 (±10)%	n/a
Low	Avg. NH_4^+ (mg N /L)	1 (±3)	3 (±5)	7 (±7)	21 (±3)	20 (±3)
	NH_4^+ Oxidation Rate (mg N/L/day)	44 (±17)	39 (±17)	25 (±17)	2 (±1)	n/a
	NH_4^+ Oxidation Efficiency (%)	93 (±14)%	84 (±23)%	62 (±34)%	3 (±2)%	n/a

The results of the 20°C study are similar to those reported in the literature. Hanaki et al. (1990) observed complete nitrification in laboratory CSTRs treating a synthetic waste with influent ammonia levels of 80 mg/L when operated at an SRT of 3.8 days. When the SRT was reduced to 1.5 days, ammonia oxidation efficiency decreased to 60%. McClintock et al. (1993) also reported decreased nitrification performance at an SRT of 1.5 days in their study using a pilot-scale activated sludge system treating domestic wastewater. The influent ammonia levels ranged from 16 mg/L to 30 mg/L, which are comparable to the levels measured in this study. McClintock et al. (1993) observed complete nitrification at SRTs ≥2.7 days, while operation at an SRT of 1.5 days resulted in only 79% nitrification. Stenstrom and Poduska (1980) also reported that nitrification could be achieved at low DO concentrations when the SRT was high, but higher DO concentrations were required for low SRT values. The results of our study support that conclusion. Nitrification in the 5-day and 2-day SRT reactors was affected by low DO to a much greater extent than in the 20- and 10-day reactors.

4.1.4.1.2 15°C Operation

Figures 4-13B through 4-16B show the influent and effluent nitrogen concentrations at each DO operational phase during the 15°C study for the 20-, 10-, 5-, and 2-day SRT reactors, respectively. As shown in these figures, the influent ammonia concentration fluctuated during this treatment study more so than during the 20°C and 10°C studies. Average ammonia concentrations and ammonia oxidation rates and efficiencies are listed in Table 4-17. Comparison of Tables 4-16 and 4-17 shows that the temperature drop from 20°C to 15°C did not appear to negatively impact ammonia oxidation rates in the 20- and 10-day SRT reactors. However, the average ammonia oxidation rates in the 5-day and 2-day reactors were significantly lower at 15°C than at 20°C.

Efficient ammonia removal occurred in the 20-day SRT reactor throughout the high and intermediate DO operating periods at 15°C (Figure 4-13B). Ammonia was present in the effluent from this reactor on only two occasions, both of which were associated with large increases in influent ammonia level. Following the decrease in DO concentration to 0.2 mg/L on August 24, 2001, ammonia was discharged from the 20-day SRT reactor several times. However, the average rate of ammonia oxidation remained high during this period. The average ammonia oxidation rate in the 20-day SRT reactor appeared to be unaffected by changes in DO concentration (Table 4-17). The ammonia oxidation rate was found to increase in all reactors during the intermediate DO level, which corresponded to an increase in the influent ammonia level.

Ammonia removal in the 10-day SRT reactor was similar to that of the 20-day reactor during operation at the high and intermediate DO levels. However, during four episodes, effluent ammonia was discharged from this reactor in response to fluctuations in influent

Table 4-17. Average (±SD) Ammonia Removal During Operation at 15°C.

DO Level	Parameter	20 day	10 day	5 day	2 day	Influent
High	Avg. NH_4^+ (mg N /L)	0.2 (±1.1)	0.4 (±1.2)	6 (±5)	12 (±6)	17 (±5)
	NH_4^+ Oxidation Rate (mg N/L/day)	43 (±17)	40 (±14)	23 (±10)	7 (±3)	n/a
	NH_4^+ Oxidation Efficiency (%)	99 (±3)%	97 (±8)%	61 (±28)%	22 (±17)%	n/a
Intermediate	Avg. NH_4^+ (mg N /L)	<0.1	0.2 (±0.9)	6 (±6)	17 (±5)	24 (±5)
	NH_4^+ Oxidation Rate (mg N/L/day)	64 (±16)	62 (±16)	41 (±21)	10 (±8)	n/a
	NH_4^+ Oxidation Efficiency (%)	100 (±0)%	98 (±11)%	65 (±32)%	16 (±12)%	n/a
Low	Avg. NH_4^+ (mg N /L)	0.2 (±0.9)	3 (±4)	11 (±4)	16 (±2)	17 (±2)
	NH_4^+ Oxidation Rate (mg N/L/day)	47 (±8)	36 (±11)	10 (±9)	0.2 (±0.5)	n/a
	NH_4^+ Oxidation Efficiency (%)	99 (±6)%	82 (±23)%	24 (±22)%	0.5 (±1.1)%	n/a

ammonia concentration (Figure 4-14B). The recovery time for the 20-day SRT reactor following a spike in the influent ammonia level was generally only 1 day; while it was generally several days for the 10-day SRT reactor. Once the DO level in the reactors was decreased to a concentration of 0.2 mg/L, ammonia removal was inconsistent in this reactor. The average ammonia oxidation rates in the 10-day SRT reactor were similar to those observed in the 20-day SRT reactor during the high and intermediate DO levels, however, during the low DO level, a slight decrease in oxidation rate was noted.

Figure 4-15B shows that ammonia treatment performance at 15°C in the 5-day SRT system was significantly poor compared to the higher SRT systems. The effluent ammonia concentration mimicked the influent concentration throughout all phases of the experiment, allowing for higher ammonia removal at lower influent concentrations. One exception occurred in the 5-day SRT reactor during the intermediate DO phase at 15°C. For approximately 26 days, ammonia oxidation was very high, even at high ammonia loading. This period corresponded to the passage of the ink-like material through the system as previously discussed. The decrease in DO from 0.5 mg/L to 0.2 mg/L greatly impacted ammonia removal in the 5-day SRT reactor, resulting in almost no ammonia oxidation during this period (Figure 4-15B). At the high and intermediate DO levels, the ammonia oxidation rate in the 5-day SRT reactor was lower than the rates observed at SRTs of 20 and 10 days by approximately 20 mg N/L/day, and this discrepancy increased at the low DO level (Table 4-17).

Nitrification performance in the 2-day SRT reactor was poor throughout the 15°C experiment, and failed completely at the low DO level. Average ammonia oxidation rates in this reactor were quite low during the high and intermediate DO levels (7 mg and 10 mg N/L/day, respectively) and approached zero at the low DO level.

The multiple linear regression model developed for the measured 15°C ammonia oxidation rates fit the experimental data quite well, as indicated by a correlation coefficient of 0.85, compared to 0.56 for the 20°C data. Seventy-three percent of the observed variability in the ammonia oxidation rate was attributed to changes in SRT, DO, and influent ammonia concentration. All of these variables were found to significantly affect ammonia oxidation rates at a 95% confidence level. Analysis of the standardized partial regression coefficients indicated that the impact of SRT on ammonia oxidation rate was greater than that of DO by a factor of two, and influent ammonia concentration by a factor of three. All three variables were found to be positively correlated. Certainly, in practice, the influent ammonia concentration would not be adjusted to increase ammonia oxidation, however, the SRT and DO levels are operational parameters that are frequently manipulated to improve process performance. The results of this linear regression analysis show that both the DO and SRT could be adjusted at 15°C to increase ammonia oxidation (up to a point). This suggests that trade-offs between the two parameters could be managed to yield maximum treatment performance.

A significant interaction between the DO level and SRT was observed in the model, and additional analyses showed that the influence of SRT on ammonia oxidation rate increased as the DO level decreased. This indicated that at the low DO level an increase in SRT would yield a greater response than at the higher DO level. DO concentration was found not to have a significant effect at an SRT of 20 days, however, for SRTs ≤10 days, the impact of DO on ammonia oxidation rate was found to increase as SRT decreased.

4.1.4.1.3 10°C Operation

Influent and effluent nitrogen concentrations are shown in Figures 4-13C through 4-16C at each DO operational phase during the 10°C study for the 20-, 10-, 5-, and 2-day SRT reactors, respectively. When the temperature was dropped to 10°C, ammonia was detected in the effluent of the 20-day SRT system during the first week, but complete nitrification was regained thereafter as shown in Figure 4-13C. In contrast, complete nitrification was not achieved in the 10-day SRT reactor at any time during the 10°C study, and steady-state (with respect to ammonia oxidation) was not reached until approximately 28 days after the temperature was dropped to 10°C. Efficient nitrification was not achieved in the 5- and 2-day SRT reactors following the decrease in temperature from 15°C to 10°C, and a substantially higher amount of ammonia was discharged from these reactors at the low temperature (Table 4-18).

The average ammonia oxidation rates in the 20-day SRT reactor during this portion of the study were similar to those observed at the higher temperatures, and DO was not shown to affect ammonia oxidation at this SRT. The average rate of ammonia oxidation incrementally decreased with decreasing SRT at all DO levels. In the 5-day SRT reactor, ammonia oxidation also

Table 4-18. Average (±SD) Ammonia Removal During Operation at 10°C.

DO Level	Parameter	20 day	10 day	5 day	2 day	Influent
High	Avg. NH_4^+ (mg N /L)	0.2 (±0.8)	6 (±6)	14 (±5)	21 (±3)	25 (±2)
	NH_4^+ Oxidation Rate (mg N/L/day)	66 (±8)	46 (±14)	21 (±11)	3 (±6)	n/a
	NH_4^+ Oxidation Efficiency (%)	99 (±3)%	74 (±23)%	37 (±19)%	5 (±9)%	n/a
Intermediate	Avg. NH_4^+ (mg N /L)	<0.1	10 (±2)	15 (±2)	21 (±2)	23 (±2)
	NH_4^+ Oxidation Rate (mg N/L/day)	63 (±9)	32 (±4)	17 (±1)	0.6 (±0.4)	n/a
	NH_4^+ Oxidation Efficiency (%)	100 (±0)%	54 (±8)%	30 (±4)%	1.0 (±0.7)%	n/a
Low	Avg. NH_4^+ (mg N /L)	0.1 (±0.5)	6 (±3)	17 (±4)	20 (±3)	21 (±4)
	NH_4^+ Oxidation Rate (mg N/L/day)	55 (±8)	36 (±4)	8 (±5)	0.2 (±0.3)	n/a
	NH_4^+ Oxidation Efficiency (%)	99 (±6)%	69 (±11)%	14 (±9)%	0.5 (±0.6)%	n/a

decreased with decreasing DO. Nitrification failure was observed in the 2-day SRT reactor throughout the low temperature study. The multiple linear regression model developed for 10°C ammonia oxidation rate data described the measured values very well (R=0.981) and showed that 96.2% of the observed variability in this parameter could be explained by changes in SRT, DO, influent ammonia, and the combined effects of SRT and DO. Both SRT and DO were found to significantly impact the ammonia oxidation rate, with SRT yielding the largest relative influence. A significant interaction between the DO and SRT was observed in the model, and in contrast to the 15°C data, the influence of SRT on ammonia oxidation rate was found to decrease slightly as the DO decreased. This may suggest that the negative impact of low DO on ammonia oxidation rate is compounded by the low temperature and cannot be counteracted by SRT as effectively as at the higher temperature.

4.1.4.2 Impact of Test Parameters on Ammonia Variability

Linear regression analysis performed on all of the data from the 20°C, 15°C, and 10°C operating periods showed that 62.3% of the observed variability in ammonia oxidation rate could be explained by changes in the SRT, temperature, DO, and influent ammonia. All four of these variables were found to significantly affect ammonia oxidation rate at a 95% confidence interval. Comparison of the magnitude of the partial regression coefficients indicated that SRT had the largest impact on ammonia oxidation rate, nearly three times that of temperature, and four times that of the influent ammonia and DO concentrations. While the temperature and influent ammonia concentration generally cannot be controlled in full-scale WWTPs, the SRT and DO can be adjusted to compensate for low temperatures. These results indicated that small improvements to the ammonia oxidation rate could be achieved by increasing the DO level. However, increasing the SRT would result in a much greater impact on nitrification performance.

The findings of this study are supported by results reported in the literature. Cheremisinoff (1996) concluded that SRT was the single most important factor in determining whether or not a wastewater treatment system will support nitrification. Reports by Hanaki et al. (1990) and McClintock et al. (1993) provide evidence to support this statement as discussed previously. Multiple linear regression analyses performed in our study showed SRT to have the greatest impact on ammonia oxidation rates at 20°C, 15°C, and 10°C, as well as when all three temperatures were considered. The results of this study are also consistent with the finding that nitrification can be achieved at low DO concentrations when the SRT is high, but higher DO concentrations were required for low SRT values (Stenstrom and Poduska, 1980). Under all temperatures in this study, ammonia oxidation in the 5-day and 2-day SRT reactors was affected by low DO to a much greater extent than the 20- and 10-day reactors.

Several researchers have demonstrated the negative impact of decreased temperature on nitrification rates and the necessity of increasing the SRT to compensate for these effects. Wild et al. (1971) concluded that it might be necessary to increase the SRT by up to five to accomplish complete nitrification at colder temperatures. McClintock et al. (1993) observed complete nitrification at an SRT of 15 days and temperatures of 20°, 15°, and 10°C, with ammonia oxidation rates of 64 mg, 80 mg, and 68 mg N/L/day, respectively. In contrast, at an SRT of 5 days, complete nitrification was observed at 20°C and 15°C, while at 10°C, only 65% conversion of ammonia-N to nitrate-N was observed. The ammonia oxidation rates measured for the 5-day SRT reactor by McClintock et al. (1993) were 80 mg, 78 mg, and 45 mg N/L/day at temperatures of 20°C, 15°C, and 10°C, respectively. Ydstebo et al. (2000) found that in order to maintain efficient nitrification it was necessary to increase the aerobic SRT in their BNR system from 4 days at 15°C to 14 days at 6°C. In our study, complete ammonia oxidation occurred at an SRT of 5 days at 20°C at the high and intermediate DO levels, while at 15°C, an SRT >5 days was required. When the temperature was dropped to 10°C, an SRT >10 days was required to achieve high rates of ammonia oxidation. Thus, the results of this study agree with the literature finding that while the rate of nitrification slows as temperature decreases, the effects can be overcome, to a considerable degree, by increasing the SRT in the activated sludge system.

4.1.4.3 Nitrite and Nitrate Data Analysis

Nitrification is a two-step process involving two separate types of bacteria that first convert ammonia to nitrite and then to nitrate. Reactor effluent ammonia, nitrite, and nitrate concentrations (as N) were used to calculate nitrite oxidation rates (equation 4.1.4.3-1) and nitrite oxidation efficiencies (equation 4.1.4.3-2).

$$NO_2^- \text{ Oxidation Rate} = \frac{Q(\text{Eff. } NO_3^-)}{V} \qquad (4.1.4.3\text{-}1)$$

where:

NO_2^- Oxidation Rate = rate of nitrite oxidation (mg N/L•day)
Q = volumetric flow rate (L/day) = 27.4 L/day
Eff. NO_3^- = effluent nitrate (mg/L as N)
V = reactor volume = 10 L

$$NO_2^- \text{ Oxidation Eff.} (\%) = \frac{(\text{Eff. } NO_3^-)}{(\text{Eff. } NO_2^- + \text{Eff. } NO_3^-)} \times 100\% \qquad (4.1.4.3\text{-}2)$$

where:

Eff. NO_2^- = effluent nitrite (mg/L as N)
Eff. NO_3^- = effluent nitrate (mg/L as N)

4.1.4.3.1 20°C Operation

High rates of nitrite oxidation were consistently achieved in the 20-, 10-, and 5-day SRT reactors during operation at the high and intermediate DO levels at 20°C (Figures 4-13A, and 4-14A, 4-15A, and Table 4-19). The presence of effluent nitrate-N concentrations comparable to influent ammonia-N levels and the lack of nitrite in the effluent of the 20- and 10-day SRT reactors indicated that complete nitrite oxidation occurred in these reactors throughout this period. In fact, the average percent conversion of the available nitrite during operation at the high and intermediate DO levels was 99% (Table 4-19). The concentration of available nitrite was computed as the sum of the reactor effluent nitrite and nitrate concentrations. While effluent nitrite averaged 0.3 mg/L in the 5-day SRT reactor during the high and intermediate DO periods of operation at 20°C, the average nitrite oxidation rates observed were roughly equivalent to those measured at SRTs of 20- and 10-days (Table 4-19). As with the ammonia oxidation rate at 20°C, when the nitrite oxidation rates measured during operation at the high and intermediate

Table 4-19. Average (±SD) Nitrite Concentrations and Nitrite Oxidation Rates During Operation at 20°C.

DO Level	Parameter	20 day	10 day	5 day	2 day
High	Avg. NO_2^- (mg N /L)	0.1 (± 0.5)	0.1 (± 0.5)	0.3 (± 0.6)	1.2 (± 0.6)
	NO_2^- Oxidation Rate (mg N/L/day)	48 (± 11)	48 (± 10)	47 (± 11)	23 (± 9)
	NO_2^- Oxidation Efficiency (%)	99 (± 7)	99 (± 7)	98 (± 7)	86 (± 8)
Intermediate	Avg. NO_2^- (mg N /L)	<0.1	<0.1	0.3 (± 0.3)	1.1 (± 0.4)
	NO_2^- Oxidation Rate (mg N/L/day)	53 (± 8)	53 (± 8)	51 (± 7)	12 (± 5)
	NO_2^- Oxidation Efficiency (%)	100 (± 0)	100 (± 0)	99 (± 1)	80 (± 7)
Low	Avg. NO_2^- (mg N /L)	0.1 (± 0.2)	0.2 (± 0.3)	3.9 (± 3.8)	0.3 (± 0.1)
	NO_2^- Oxidation Rate (mg N/L/day)	48 (± 11)	42 (± 12)	17 (± 14)	1.1 (± 0.6)
	NO_2^- Oxidation Efficiency (%)	99 (± 2)	98 (± 3)	61 (± 25)	52 (± 8)

DO levels in the 20-, 10-, and 5-day SRT reactors were included in a multiple regression analysis, only the concentration of available nitrite was found to significantly affect nitrite oxidation rates at a 95% confidence level. This result suggests that at an operating temperature of 20°C, increasing the DO concentration above 1.0 mg/L or the SRT above 5 days does not enhance nitrification performance.

Nitrite oxidation efficiency in the 20-day SRT reactor remained high during operation at a DO concentration of 0.5 mg/L, but was initially impacted by the change in DO level from 0.5 to 0.2 mg/L on November 29, 2000. Complete nitrite oxidation was eventually recovered in this reactor (Figure 4-13A), and the average nitrite oxidation rate during the low DO operating period was consistent with the average nitrite oxidation rates measured at the high and intermediate DO operating periods. Although the walls of the reactors were scrubbed on a daily basis to inhibit attached growth, biofilm development cannot be completely discounted as a contributing factor in nitrification at DO levels of 0.2 mg/L. Nitrite oxidation efficiency also remained high in the 10-day SRT reactor during operation at a DO concentration of 0.5 mg/L. However, in contrast to the 20-day SRT reactor, complete nitrification was not achieved in the 10-day SRT reactor during operation at 0.2 mg/L (Figure 4-14A).

Nitrite oxidation in the 5-day SRT reactor was strongly impacted by the decrease in DO to the low level (Figure 4-15A), and a large decline in average nitrite oxidation rate was noted during this operating period. The decrease in nitrite oxidation rate resulted not only from a decline in available nitrite due to incomplete ammonia oxidation (Table 4-16) but also from a decline in nitrite oxidation efficiency (Table 4-19). Although ammonia oxidation has been reported to be the rate-limiting step in nitrification, Figure 4-15A indicates a build-up of nitrite in the effluent of the 5-day SRT reactor during 0.5 mg/L DO operation. It was reported by Laanbroek et al. (1994) that ammonia-oxidizers had a higher affinity for oxygen (i.e., a lower half-saturation constant) than nitrite-oxidizers. Therefore, NOB may have difficulty competing with AOB for available oxygen. A downward trend in effluent nitrite over time suggests that the nitrite-oxidizers began to acclimate to the low DO conditions.

Multiple linear regression analyses were performed individually for each SRT with nitrite oxidation rates measured during the intermediate and low DO level operating periods at 20ºC. For the 2-day SRT reactor, both DO level and available nitrite concentration significantly impacted oxidation rate at a 95% confidence interval, with nitrite concentration yielding 10 times as much influence as the DO concentration. For the 5-day SRT reactor, only the DO concentration significantly impacted the nitrite oxidation rate, which is supported by the fact that a nitrite build-up occurred. This indicates that for a time, nitrite oxidation controlled the rate of nitrification in this reactor, which was not observed at the other SRTs. According to multiple linear regression analysis of the 10-day SRT data, both DO and available nitrite concentration significantly impacted the oxidation rate, with the available nitrite concentration having nearly 30 times the impact of DO concentration. For the 20-day SRT reactor, only the amount of available nitrite significantly impacted the nitrite oxidation rate under these operating conditions.

The nitrogen speciation data for the 2-day SRT reactor, shown in Figure 4-16A for the 20°C study, were much different than that of the other reactors at this temperature even under excess DO conditions. During the high and intermediate DO operating periods, significant ammonia and nitrite concentrations were regularly measured in the effluent of the 2-day SRT reactor and the average nitrite oxidation rates were much lower than those measured at higher SRTs. At the high DO level, the average nitrite oxidation rate was approximately 50% lower than observed for the 20-, 10-, and 5-day SRTs. However, the percentage of available nitrite oxidized was only 12% lower than the higher SRTs. Similarly, at the intermediate DO level, the average nitrite oxidation rate and efficiency were 75% and 18% lower than at the higher SRTs, respectively. This data suggests that the decline in nitrite oxidation rate was caused in large part by the lowered amount of available nitrite. Operation of the 2-day SRT reactor at the low DO level resulted in near complete nitrification failure. Data from January 2 to January 10 shows that ammonia oxidation had essentially ceased. For these dates, only trace amounts of nitrate and nitrite (averaging 0.4 mg/L each) could be measured in the effluent. This data again shows much higher nitrite oxidation efficiency than ammonia oxidation efficiency (52% versus 3%, respectively) supporting the theory that ammonia oxidation is the rate-limiting step. Multiple linear regression analysis of the 2-day SRT reactor suggested that the concentration of available nitrite controlled the nitrite oxidation rate rather than the DO concentration at this SRT.

4.1.4.3.2 15°C Operation

Table 4-20 provides a tabulation of the average effluent nitrite concentrations and nitrite oxidation rates and efficiencies measured during each DO level in each reactor during the 15°C study. Comparison of Table 4-19 to Table 4-20 shows that the average nitrite oxidation rates in the 20-day SRT reactors were similar, suggesting that temperature did not significantly affect nitrification under these conditions. Nitrite oxidation rates in the 20-day SRT also did not appear to be negatively impacted by decreasing DO at this temperature. Nitrification was initially disrupted in the 20-day SRT reactor when the DO concentration was dropped to 0.2 mg/L. However, complete nitrification was eventually restored in this reactor (Figure 4-13B). Multiple linear regression of the measured nitrite oxidation rates in the 20-day SRT reactor at 15°C showed that only available nitrite concentration significantly impacted the rate of nitrite oxidation at this SRT.

Nitrite oxidation efficiency in the 10-day SRT reactor was similar to that observed in the 20-day SRT reactor during the high and intermediate DO levels. However, there were a few more instances where effluent ammonia and nitrite were measured in the 10-day SRT reactor

Table 4-20. Average (±SD) Nitrite Concentrations and Nitrite Oxidation Rates During Operation at 15°C.

DO Level	Parameter	20 day	10 day	5 day	2 day
High	Avg. NO_2^- (mg N /L)	0.2 (± 0.3)	0.3 (± 0.4)	0.6 (± 0.3)	0.4 (± 0.1)
	NO_2^- Oxidation Rate (mg N/L/day)	43 (± 17)	39 (± 14)	21 (± 10)	6 (± 3)
	NO_2^- Oxidation Efficiency (%)	99 (± 4)	98 (± 1)	93 (± 5)	81 (± 12)
Intermediate	Avg. NO_2^- (mg N /L)	0.1 (± 0.1)	0.3 (± 0.5)	0.9 (± 0.5)	0.6 (± 0.3)
	NO_2^- Oxidation Rate (mg N/L/day)	64 (± 16)	62 (± 16)	38 (± 20)	8 (± 8)
	NO_2^- Oxidation Efficiency (%)	99 (± 1)	98 (± 3)	92 (± 5)	78 (± 12)
Low	Avg. NO_2^- (mg N /L)	0.2 (± 0.1)	0.2 (± 0.2)	0.4 (± 0.2)	<0.1
	NO_2^- Oxidation Rate (mg N/L/day)	47 (± 8)	35 (± 11)	9 (± 9)	0.1 (± 0.4)
	NO_2^- Oxidation Efficiency (%)	99 (± 1)	98 (± 1)	86 (± 9)	73 (± 38)

(Figure 4-14B). A decrease in average nitrite oxidation rate was noted at the low DO level. However, the percentage of available nitrite oxidized remained at 98% (Table 4-20), suggesting the decline in the oxidation rate was the result of incomplete ammonia oxidation (Table 4-17). Multiple linear regression analysis of the 10-day SRT data showed that only the concentration of available nitrite significantly affected nitrite oxidation rates at a 95% confidence level. Nitrite oxidation rates in the 5-day SRT reactor were significantly lower at 15°C than at 20°C and were drastically reduced at the low DO level. Even though the oxidation rates were lower, the percent nitrite oxidation efficiency remained >85% in the 5-day SRT reactor throughout operation at 15°C, suggesting that the reduction in rate was primarily due to the decrease in available nitrite. As shown in Figure 4-16B, the 2-day SRT reactor failed to achieve efficient nitrification during any phase of this experiment, and nitrification failed completely during the low DO phase at 15°C. Even so, in excess of 70% of the available nitrite was converted to nitrate during 15°C operation, suggesting that nitrification was limited by the ammonia oxidation step. According to multiple linear regression analyses, only available nitrite concentration was found to significantly impact the nitrite oxidation rate at a 95% confidence level at SRTs of both 5 and 2 days.

The multiple linear regression model developed for the 15°C nitrite oxidation rate data described the measured rates very well (R=1.0) and indicated that 99.9% of the observed variability in nitrite oxidation rate could be attributed to changes in concentration of available nitrite, SRT, and DO levels. All of the independent variables included in the model significantly affected the nitrite oxidation rate, with available nitrite concentration exerting nearly three times as much influence on the nitrite oxidation rate as SRT and nearly four times the influence of DO concentration.

4.1.4.3.3 10°C Operation

Efficient nitrite oxidation occurred in the 20-day SRT reactor throughout all phases of the 10°C experiment (Figure 4-13C), suggesting that the temperature drop and changes in DO did not significantly impact nitrification at this SRT. In contrast, nitrite oxidation in the 10-day SRT reactor was severely impacted by the decrease in temperature. Figure 4-14C shows a build-up of nitrite in the effluent following the temperature change, similar to the build-up observed in the 5-day SRT reactor during operation at the low DO level at 20°C (Figure 4-15A). Randall and Buth (1984) noted a similar phenomenon in laboratory- and full-scale activated sludge systems and suggested that although the maximum possible rate of nitrate formation may be substantially greater than the maximum rate of nitrite formation at 20°C, there exists a lower critical temperature below which the rate of nitrite formation will exceed that of nitrate, resulting in a build-up of nitrite. In their experiments, the critical temperatures for a laboratory-scale and a full-scale reactor were 14°C and 12°C, respectively. Below the critical temperature, the rate of nitrate formation controlled the overall nitrification rate. In our study, the critical temperature for the 10-day SRT reactor appears to be between 15°C and 10°C. After an extended transition period, the effluent nitrite levels decreased until this substance was no longer consistently detected in the 10-day SRT reactor effluent. A corresponding increase in the average conversion efficiency of nitrite to nitrate was observed. These results suggest that the nitrite oxidizers were able to acclimate to the low temperature conditions. The ammonia oxidizers did not appear to fully acclimate to the low temperature as evidenced by the significant amount of ammonia-N measured in the effluent of this reactor throughout the remainder of the 10°C study (Figure 4-14C). Even though nitrite was no longer routinely present in the effluent from the 10-day SRT reactor after the acclimation period, nitrite oxidation rates remained suppressed during the intermediate and low DO portions of the study (Table 4-21) because of the simultaneous decline in ammonia oxidation rate.

Figure 4-15C shows that nitrification was inhibited in the 5-day SRT reactor when operated at a temperature of 10°C. Nitrification performance decreased with decreasing DO levels, and complete nitrification failure in the 5-day SRT system was documented during the final days of the low DO phase at 10°C (as evidenced by the very low effluent nitrite and nitrate concentrations). While the average nitrite oxidation rate decreased from 20 mg N/L/day to 7 mg N/L/day from the high to low DO levels, the nitrite oxidation efficiency only decreased from 92% to 88%. As shown in Figure 4-16C, the 2-day SRT reactor remained at a state of complete failure throughout the duration of the 10°C experiment. Even so, >75% of the available nitrite was converted to nitrite, suggesting once again that incomplete ammonia oxidation was responsible for the decrease in nitrite oxidation rate.

Multiple linear regression analysis of the 10°C data showed that available nitrite concentration and SRT significantly impacted the nitrite oxidation rate at a 95% confidence level, with the available nitrite concentration yielding more than 20 times more influence on the rate than SRT. Further analysis revealed that at the high DO level, only the available nitrite concentration influenced the nitrite oxidation rate, whereas at the intermediate and low DO levels both available nitrite and SRT were significant.

Table 4-21. Average (±SD) Nitrite Concentrations and Nitrite Oxidation Rates During Operation at 10°C.

DO Level	Parameter	20 day	10 day	5 day	2 day
High	Avg. NO_2^- (mg N /L)	0.2 (± 0.1)	2.4 (± 3.0)	0.6 (± 0.3)	0.1 (± 0.1)
	NO_2^- Oxidation Rate (mg N/L/day)	66 (± 7)	40 (± 14)	20 (± 10)	3 (± 6)
	NO_2^- Oxidation Efficiency (%)	99 (± 0)	87 (± 16)	92 (± 2)	90 (± 14)
Intermediate	Avg. NO_2^- (mg N /L)	0.5 (± 1.2)	0.5 (± 0.1)	0.5 (± 0.0)	0.1 (± 0.1)
	NO_2^- Oxidation Rate (mg N/L/day)	62 (± 11)	30 (± 4)	15 (± 1)	0.3 (± 0.2)
	NO_2^- Oxidation Efficiency (%)	97 (± 7)	96 (± 1)	91 (± 0)	73 (± 28)
Low	Avg. NO_2^- (mg N /L)	0.1 (± 0.1)	0.3 (± 0.1)	0.3 (± 0.1)	<0.1
	NO_2^- Oxidation Rate (mg N/L/day)	55 (± 8)	35 (± 4)	7 (± 5)	0.1 (± 0.1)
	NO_2^- Oxidation Efficiency (%)	100 (± 1)	98 (± 1)	88 (± 7)	76 (± 34)

4.1.4.4 Impact of Test Parameters on Nitrite Variability

When all operational conditions were included in the linear regression model, 94% of the observed variability in nitrite oxidation rate could be explained by changes in SRT, temperature, DO concentration, and available nitrite concentration. Of the four variables, only temperature did not significantly impact nitrite oxidation at the 95% confidence level. The relative impact of available nitrite concentration on nitrite oxidation was found to be five times greater than SRT, and more than 15 times greater than DO.

In general, the percent of nitrite oxidized was lower for the 5- and 2-day SRTs than for the 20- and 10-day SRTs. For the lower SRTs, the percent nitrite oxidized decreased with decreasing DO. However, the data suggests that nitrite oxidation rate was controlled by the amount of nitrite available, which was controlled by the ammonia oxidation rate. Since ammonia oxidation was found to be the rate-limiting step, the impact of SRT, DO, and temperature on the ammonia oxidation rate indirectly affected the nitrite oxidation rate.

4.1.4.5 Alkalinity

4.1.4.5.1 20°C Operation

Influent and effluent alkalinity measurements were taken to ensure that sufficient buffer capacity existed to avoid a major depression of pH. Values are shown in Figure 4-17 for the 20°C study over time. The 2-day SRT reactor regularly discharged a higher concentration of alkalinity than the other reactors. This reinforces that the 2-day SRT reactor did not achieve complete nitrification over this study. Average alkalinities for the influent and reactor effluents are reported in Table 4-22 for the 20°C study.

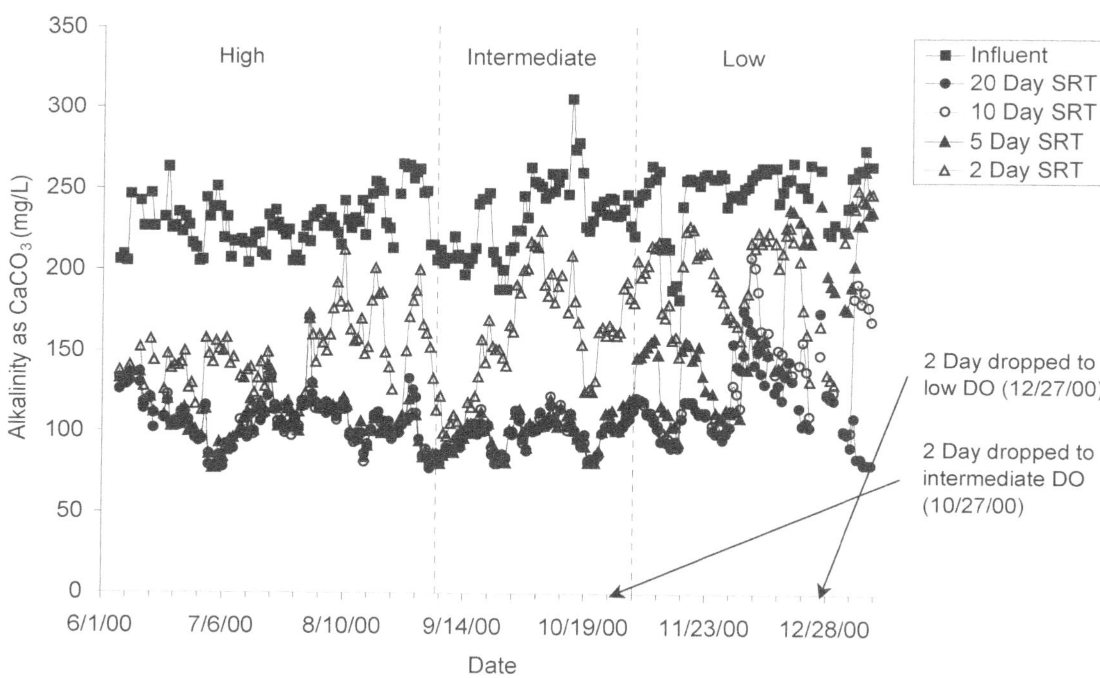

Figure 4-17. Fluctuations in Effluent Alkalinity as a Function of DO Concentration and SRT During 20°C Operation.

Table 4-22. Average (±SD) Alkalinity (mg/L as $CaCO_3$) During 20°C Operation.

DO Level	High	Intermediate	Low
Influent	229 (±16)	232 (±23)	247 (±19)
20-Day SRT	106 (±16)	98 (±10)	116 (±23)
10-Day SRT	107 (±16)	99 (±10)	131 (±31)
5-Day SRT	108 (±16)	100 (±11)	161 (±44)
2-Day SRT	150 (±29)	193 (±24)	214 (±50)

The alkalinity measurements, along with effluent nitrate and nitrite concentrations, were also used to verify the stoichiometric ratio of 7.1 mg alkalinity (as $CaCO_3$) consumed per milligram of ammonia (as N) converted to nitrate. The difference in influent and effluent alkalinity divided by the sum of effluent nitrate and nitrite was used to compute the ratio. Figure 4-18 shows that the ratio was upheld for the duration of the treatment study in the 20- and 10-day SRT reactors. The 5-day SRT reactor maintained the ratio until this system began to fail with respect to nitrification at a DO of 0.2 mg/L. The 2-day reactor showed similar results during the 0.5 mg/L DO operational period.

4.1.4.5.2 15°C Operation

Figure 4-19 shows influent and effluent alkalinity values for the 15°C study and average values for each DO operational phase are presented in Table 4-23. Again the 2-day SRT reactor

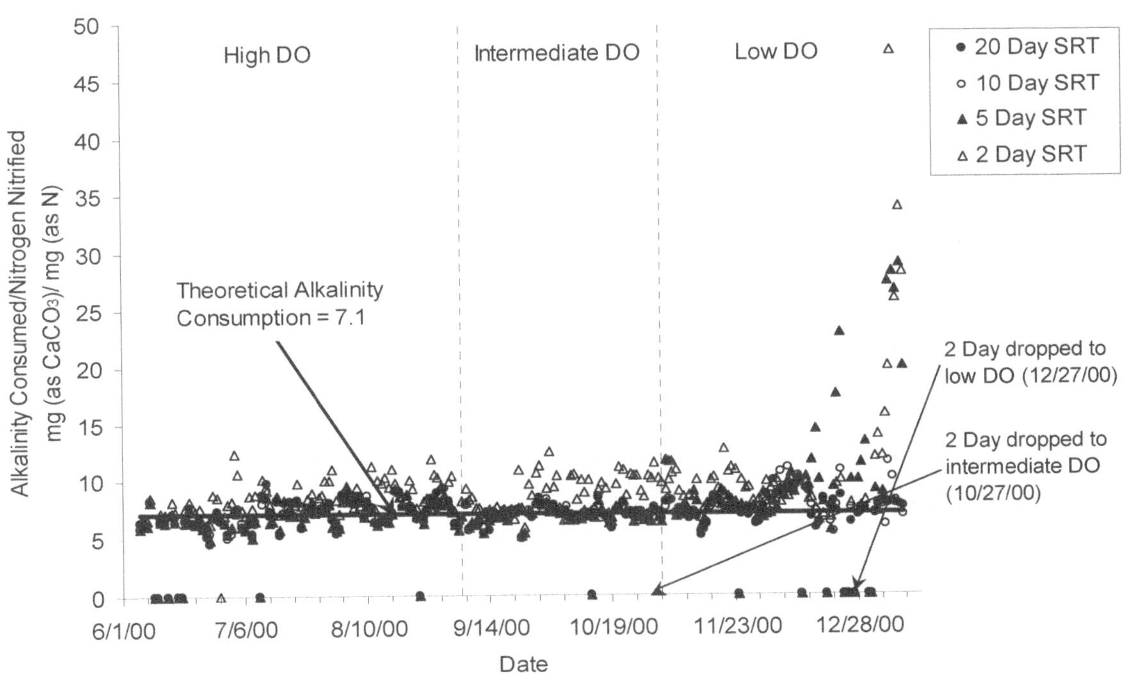

Figure 4-18. Change in Alkalinity Consumption Ratio Due to Inconsistency of Nitrification at Low DO During 20°C Operation for Each SRT.

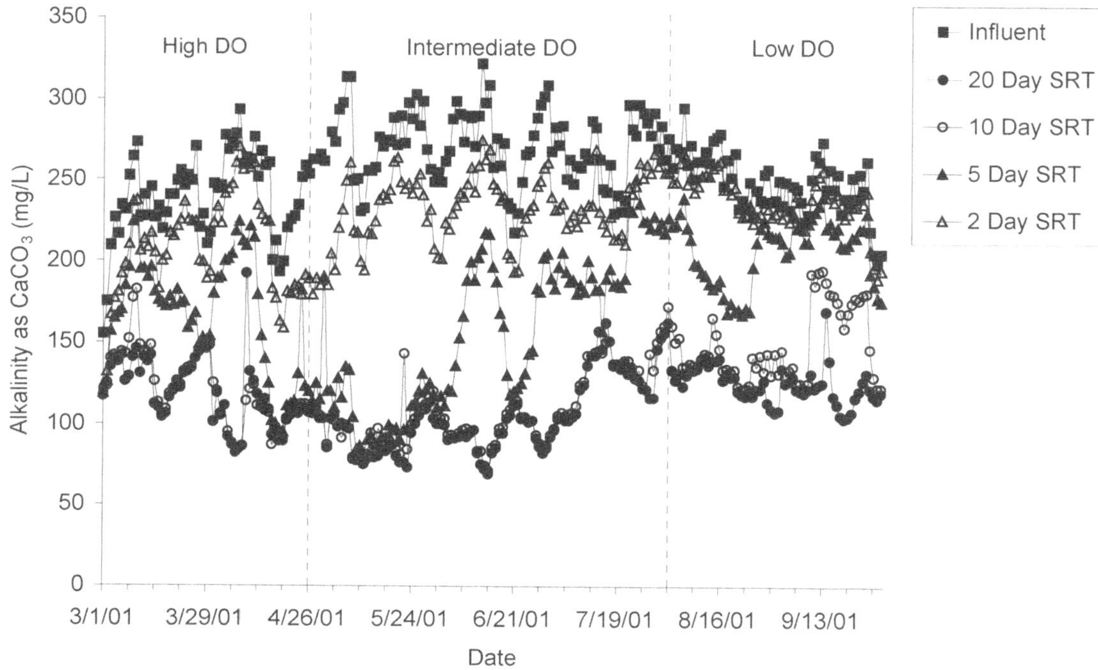

Figure 4-19. Fluctuations in Effluent Alkalinity as a Function of DO Concentration and SRT During 15°C Operation.

Table 4-23. Average (±SD) Alkalinity (mg/L as $CaCO_3$) During 15°C Operation.

DO Level	High	Intermediate	Low
Influent	239 (±27)	271 (±22)	248 (±19)
20-Day SRT	120 (±21)	104 (±22)	125 (±12)
10-Day SRT	122 (±23)	107 (±22)	148 (±22)
5-Day SRT	168 (±36)	155 (±46)	206 (±20)
2-Day SRT	209 (±32)	231 (±22)	236 (±16)

regularly discharged a substantially higher alkalinity level than the other reactors. Unlike the 20°C study, during 15°C operation the 5-day SRT reactor also consistently discharged higher alkalinity levels than the 10- and 20- day SRT reactors. The increase in effluent alkalinity corresponds to the decrease in ammonia oxidation that occurred during this period.

Figure 4-20 shows the change in alkalinity consumption ratio over the 15°C operational period. The ratio was upheld for the duration of the study in the 20- and 10-day SRT reactors. The 5-day SRT reactor maintained the ratio, for the most part, until it began to fail, with respect to nitrification, during the low DO period. Figure 4-20 shows a much greater variability in this ratio for the 2-day SRT reactor during the entire 15°C study when compared to the 20°C study.

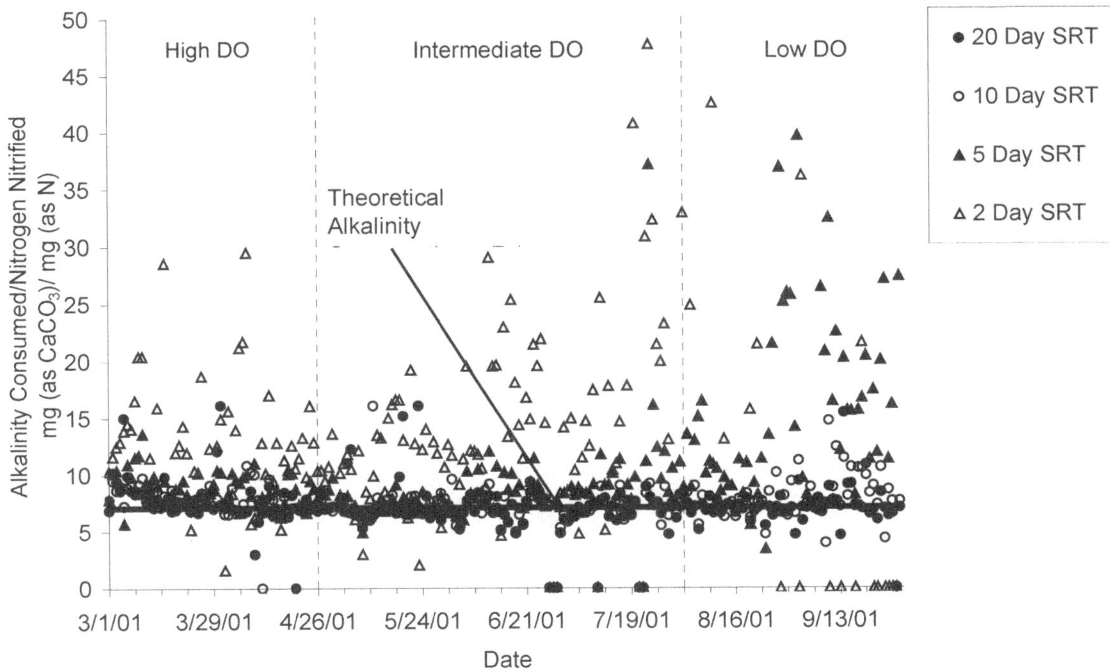

Figure 4-20. Change in Alkalinity Consumption Ratio Due to Inconsistent Nitrification at Low DO During 15°C Operation for each SRT.

4.1.4.5.3 10°C Operation

Influent and effluent alkalinity values for the 10°C study are shown in Figure 4-21. Table 4-24 presents average values for each DO operational phase. During this study, a trend of increasing effluent alkalinity with decreasing SRT can be noted, which was not observed at the higher temperatures.

Figure 4-22 shows the change in the alkalinity consumption ratio over the 10°C operational period. Once again the ratio was upheld for the duration of the treatment study in the 20- and 10-day SRT reactors. The 5-day SRT reactor showed more variability in the alkalinity consumption ratio than at higher temperatures and again increased sharply during the low DO period, signaling nitrification failure. As in the 15°C study, the 2-day SRT reactor failed to maintain the ratio at all DO levels.

Table 4-24. Average (±SD) Alkalinity (mg/L as $CaCO_3$) During 10°C Operation.

DO Level	High	Intermediate	Low
Influent	259 (±14)	247 (±9)	254 (±14)
20-Day SRT	89 (±12)	86 (±11)	114 (±13)
10-Day SRT	132 (±33)	150 (±8)	152 (±14)
5-Day SRT	187 (±10)	187 (±10)	223 (±19)
2-Day SRT	230 (±10)	230 (±10)	245 (±12)

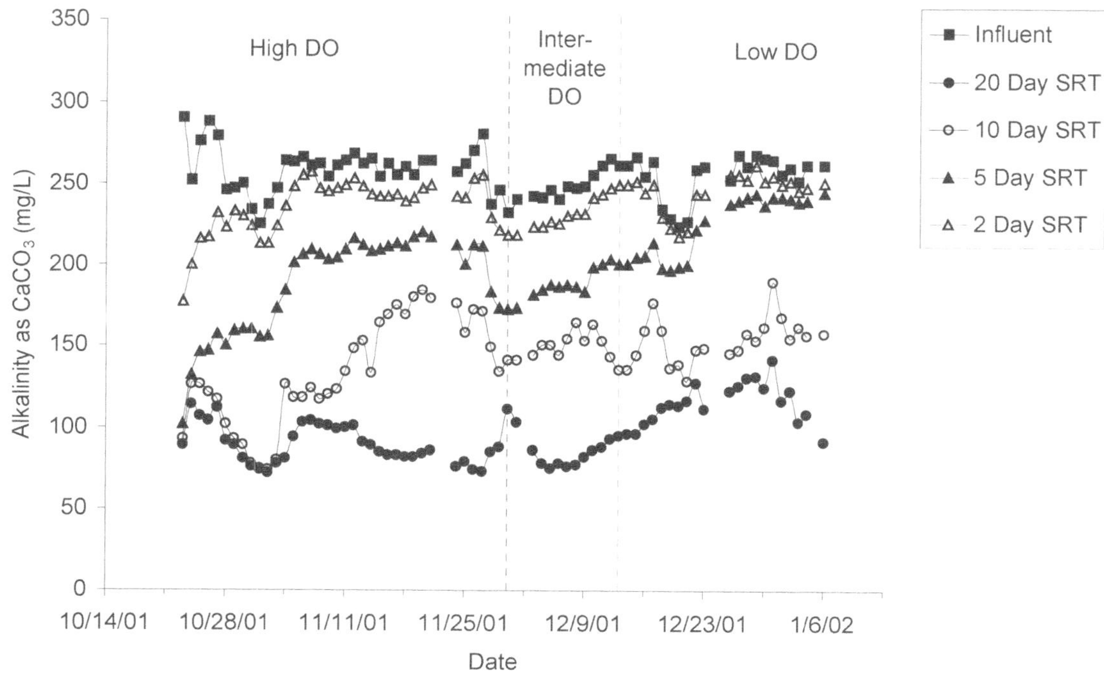

Figure 4-21. Fluctuations in Effluent Alkalinity as a Function of DO Concentration and SRT During 10°C Operation.

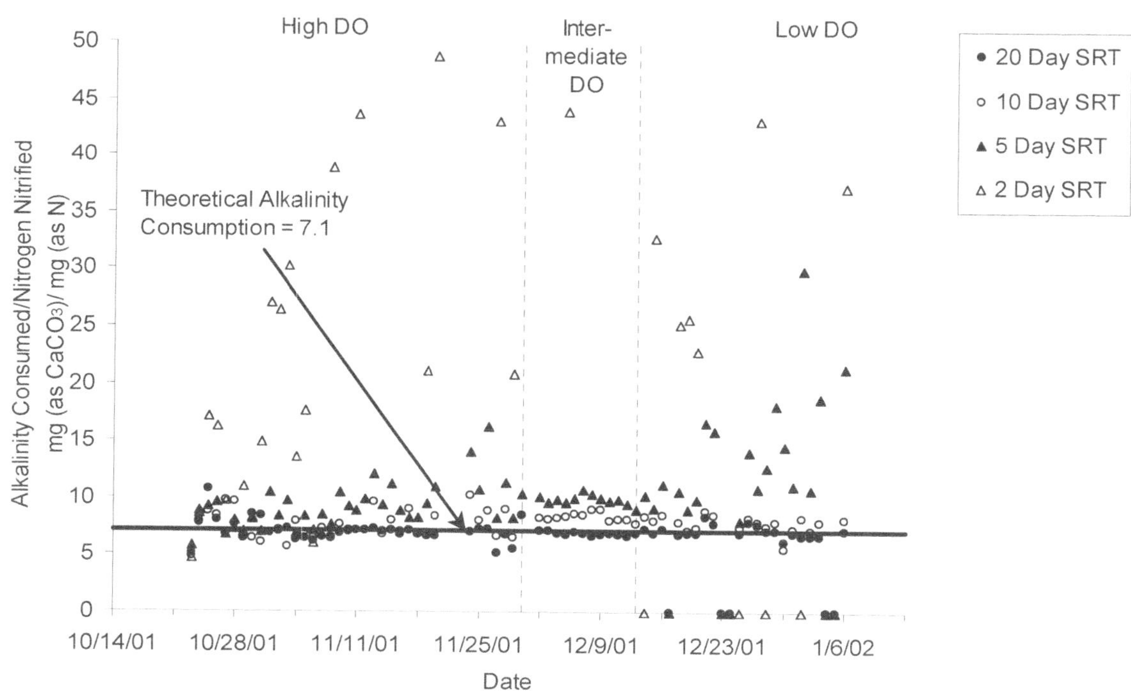

Figure 4-22. Change in Alkalinity Consumption Ratio Due to Inconsistent Nitrification at Low DO During 10°C Operation for Each SRT.

4.1.5 Molecular Analysis and Comparison to Traditional Analysis

The overall goal of this research was to quantify specific N-utilizing bacteria under varying operational conditions during activated sludge treatment using molecular based techniques. Total bacteria, ammonia-oxidizing bacteria, and nitrite-oxidizing bacteria were monitored using a suite of real-time PCR assays. For the municipal bench scale reactor system, three real-time PCR assays were designed and optimized for three targets, bacterial 16S rDNA, *Nitrospira* 16S rDNA, and *N. oligotropha*-type *amo*A.

4.1.5.1 Selection of Samples for Molecular Analysis

An experimental matrix (Table 4-25) was prepared to develop a sample analysis schedule. The data were grouped into sets based on operating conditions (i.e., temperature, DO level, and SRT). For analysis purposes, the DO levels were collapsed into 3 groups: high (DO = 2.0 mg/L and 3.0 mg/L), intermediate (DO = 1.5 mg/L and 1.0 mg/L) and low (DO = 0.5 mg/L and 0.2 mg/L). Because it was not feasible to perform real-time PCR analysis on all reactor samples (collected daily throughout the study), it was necessary to select subsets of samples representative of each experimental phase. Listings of the selected sample dates for each SRT and DO level are shown in Tables 4-26, 4-27, and 4-28 for the 20°C, 15°C, and 10°C systems, respectively. Sample dates denoted in bold indicate apparent steady-state samples. Steady-state periods were defined as time intervals in which stable nitrification occurred after the reactors were operated for a minimum of twice the SRT following a change in environmental or operational conditions.

4.1.5.2 Total Eubacterial Concentration

In addition to MLVSS, total eubacterial concentration determined from real-time PCR analysis was used as an overall measure of biomass content in each reactor. Real-time PCR data was measured in terms of the number of gene copies in a sample. To allow for a better comparison with the traditional information collected in this study, the real-time PCR data was converted to a cell per liter (mixed liquor) basis. The conversion of total 16S rDNA copy number to cell number can be problematic since bacterial ribosomal operons can vary from 1 to 15 (Condon, 1995; Klappenbach et al., 2000). It has been hypothesized that bacteria with faster growth rates or the ability to respond rapidly to environmental changes have higher ribosomal operon numbers than slow growing bacteria that are sensitive to environmental changes

Table 4-25. Municipal Bench Scale Experimental Matrix.

	Temperature			
DO (mg/L)	20°C	15°C	10°C	5°C
Excess	X		X	X
3.0	X	X		
2.0	X	X	X	
1.5	X	X		
1.0	X	X	X	
0.5	X	X	X	
0.2	X	X		

Table 4-26. 20°C Municipal Bench-Scale System Sample Dates* for Molecular Analysis.

D.O. (mg/L)	SRT 2 day	SRT 5 day	SRT 10 day	SRT 20 day
High	**07/28/00**	**07/28/00**	08/04/00	**08/04/00**
	08/03/00	08/03/00	**08/13/00**	**08/13/00**
	08/20/00	08/04/00	**08/20/00**	**08/20/00**
	09/01/00	**08/16/00**	**08/21/00**	**08/21/00**
	09/04/00	**08/21/00**	**09/01/00**	09/01/00
	10/05/00		**09/04/00**	**09/04/00**
	10/19/00			
Intermediate	**12/05/00**	10/05/00	10/05/00	10/05/00
	12/06/00	**10/19/00**	**10/19/00**	10/19/00
	12/07/00	**10/22/00**	**10/22/00**	**10/22/00**
	12/11/00	**10/24/00**	**10/24/00**	**10/24/00**
		10/26/00	**10/26/00**	**10/26/00**
Low	**01/03/01**	11/08/00	11/08/00	11/08/00
	01/05/01	11/16/00	11/16/00	11/16/00
	01/06/01	11/28/00	11/28/00	11/28/00
	01/07/01	**12/01/00**	12/01/00	12/01/00
	01/09/01	**12/05/00**	12/05/00	12/05/00
		12/06/00	12/06/00	12/06/00
		12/07/00	12/07/00	12/07/00
		12/11/00	12/11/00	12/11/00

*Dates in bold denote steady-state samples

Table 4-27. 15°C Municipal Bench-Scale System Sample Dates* for Molecular Analysis.

D.O. (mg/L)	SRT 2 day	SRT 5 day	SRT 10 day	SRT 20 day
High	**03/06/01**	03/06/01	03/06/01	03/06/01
	03/14/01	03/14/01	03/14/01	03/14/01
	04/03/01	**04/03/01**	**04/03/01**	04/03/01
	04/18/01	**04/18/01**	**04/18/01**	**04/18/01**
	04/23/01	**04/23/01**	**04/23/01**	**04/23/01**
Intermediate	**07/14/01**	**07/14/01**	**07/14/01**	**07/14/01**
	07/16/01	**07/16/01**	**07/16/01**	**07/16/01**
	07/25/01	**07/25/01**	**07/25/01**	**07/25/01**
	07/27/01	**07/27/01**	**07/27/01**	**07/27/01**
	07/29/01	**07/29/01**	**07/29/01**	**07/29/01**
Low	**09/11/01**	**09/11/01**	**09/11/01**	**09/11/01**
	09/13/01	**09/13/01**	**09/13/01**	**09/13/01**
	09/18/01	**09/18/01**	**09/18/01**	**09/18/01**
	09/21/01	**09/21/01**	**09/21/01**	**09/21/01**
	09/24/01	**09/24/01**	**09/24/01**	**09/24/01**

*Dates in bold denote steady-state samples

Table 4-28. 10°C Municipal Bench-Scale System Sample Dates* for Molecular Analysis.

	SRT			
D.O. (mg/L)	2 day	5 day	10 day	20 day
High	**11/18/01**	**11/18/01**	**11/18/01**	11/18/01
	11/20/01	**11/20/01**	**11/20/01**	11/20/01
	11/21/01	**11/21/01**	**11/21/01**	11/21/01
	11/24/01	**11/24/01**	**11/24/01**	11/24/01
	11/27/01	**11/27/01**	**11/27/01**	11/27/01
Intermediate	**12/03/01**	12/03/01	12/03/01	12/03/01
	12/05/01	12/05/01	12/05/01	12/05/01
	12/07/01	12/07/01	12/07/01	12/07/01
	12/09/01	**12/09/01**	12/09/01	12/09/01
	12/12/01	**12/12/01**	12/12/01	12/12/01
Low	**12/23/01**	**12/23/01**	12/23/01	12/23/01
	12/26/01	**12/26/01**	12/26/01	12/26/01
	12/27/01	**12/27/01**	12/27/01	12/27/01
	01/02/02	**01/02/02**	01/02/02	01/02/02
	01/04/02	**01/04/02**	01/04/02	01/04/02

*Dates in bold denote steady-state samples

(Condon, 1995; Klappenbach et al., 2000). Therefore, it is possible that the low SRT reactor may select for microbial populations with a different number of ribosomal operons. The reported ribosomal operon copy numbers for bacterial species found in wastewater treatment systems range from a high of 13 in *Bacillus* to 1 in *Nitrosomonas* (Klappenbach et al., 2001). Given that the majority of species in wastewater treatment reactors contain a relatively low number of ribosomal operons, the current average operon number for all bacterial species of 3.6 may be the most reasonable conversion factor (Klappenbach et al., 2001). Therefore, a value of 3.6 16S rDNA gene copies per cell was assumed in this study for the calculation of total eubacterial cellular concentrations.

4.1.5.2.1 20°C Operation

Table 4-29 lists the average total eubacterial 16S rDNA cell concentrations normalized to mixed liquor volume for the 20°C system. The eubacterial cell densities reported in this table are consistent with those reported in several studies in the literature. Harms et al. (2003) reported an average value of 4.3 (±2.0) x 10^{11} eubacterial cells/L when applying the same real-time PCR assay to 12 monthly samples collected from the full-scale municipal WWTP originally used to seed the bench-scale reactors. The WWTP operates at an average SRT of 10 days. Lishman et al.

Table 4-29. Average (±SD) Total Eubacterial Concentrations (cells/L) During Operation at 20°C.

DO Level	High	Intermediate	Low
20-Day SRT	11.0 (±0.5) x 10^{11}	6.2 (±1.6) x 10^{11}	7.1 (±4.1) x 10^{11}
10-Day SRT	7.0 (±3.7) x 10^{11}	5.2 (±0.5) x 10^{11}	6.5 (±3.8) x 10^{11}
5-Day SRT	3.8 (±2.9) x 10^{11}	2.0 (±1.2) x 10^{11}	1.8 (±0.8) x 10^{11}
2-Day SRT	1.3 (±0.3) x 10^{11}	0.75 (±4.5) x 10^{11}	1.1 (±0.2) x 10^{11}

(2000) obtained similar results in a sequencing batch reactor (SBR) operated at a 10-day SRT and temperatures of 19°C and 14°C. In their study, the total number of bacteria larger than 0.2 μm, enumerated using acridine orange direct counts (AODC), were between 4.9×10^{11} and 1.2×10^{12} cells/L. Lee et al. (2002) reported biomass densities of 2.32×10^{11} to 6.12×10^{11} cells/L for a bench-scale activated sludge reactor operated at an SRT of 10 days, a temperature of 20°C, and excess DO. The biomass densities in their study were determined by performing colony counts.

Table 4-29 shows that total eubacterial concentrations follow the expected trends with the highest cell concentrations present in the 20-day SRT samples and the lowest cell concentrations in the 2-day SRT samples. While total eubacterial concentration (cells/L) was highly dependent on SRT, significant trends in bacterial concentration as a function of DO level were not observed. Figure 4-23 provides a comparison the average total eubacterial and MLVSS concentrations as a function of SRT and DO level and shows that both measures of biomass followed similar trends with respect to SRT.

Multiple linear regression analysis was performed on the total eubacterial concentrations measured during 20°C operation in the same manner as that used for MLVSS. This model did not describe the data as well as the model developed for MLVSS. The correlation coefficient was 0.769 for total eubacterial concentration, compared to 0.926 for MLVSS. Only 59% of the variability in measured eubacterial concentrations could be explained by changes in controlled variables, in contrast to 86% for the MLVSS. As in the multiple linear regression model for MLVSS, of the four independent variables included in the analysis (SRT, DO, influent COD, and joint effect of SRT and DO) only the SRT was found to significantly affect the concentration of total eubacteria in the reactors.

4.1.5.2.2 15°C Operation

The average total eubacterial concentrations measured during 15°C operation are listed in Table 4-30. Again the highest cell concentrations were measured in the 20-day SRT samples and the lowest cell concentrations were present in the 2-day SRT samples. Comparison of Tables 4-29 and 4-30 shows that a trend in concentration of total eubacteria based on the decrease in temperature from 20°C to 15°C was not observed. Figure 4-24 compares the average MLVSS and total bacteria concentrations in the reactor during the 15°C study and shows a similar trend in response to changes in SRT. Average values of both measures of biomass content are shown to increase between the high and intermediate DO levels and then remain fairly stable when the DO is dropped to the low level.

The regression model developed for the 15°C total eubacteria data was similar to that obtained for the 20°C data (R=0.782). This model revealed that 61.1% of the observed variability in cell concentrations could be attributed to changes in the influent COD, SRT, DO, and the

Table 4-30. Average (±SD) Total Eubacterial Concentrations (cells/L) During Operation at 15°C.

DO Level	High	Intermediate	Low
20-Day SRT	$3.7 (\pm 2.0) \times 10^{11}$	$9.8 (\pm 1.5) \times 10^{11}$	$16.0 (\pm 0.3) \times 10^{11}$
10-Day SRT	$3.7 (\pm 1.5) \times 10^{11}$	$7.8 (\pm 2.0) \times 10^{11}$	$16.0 (\pm 0.1) \times 10^{11}$
5-Day SRT	$3.1 (\pm 1.9) \times 10^{11}$	$8.0 (\pm 2.6) \times 10^{11}$	$5.7 (\pm 0.7) \times 10^{11}$
2-Day SRT	$1.5 (\pm 0.7) \times 10^{11}$	$4.3 (\pm 0.5) \times 10^{11}$	$2.6 (\pm 0.6) \times 10^{11}$

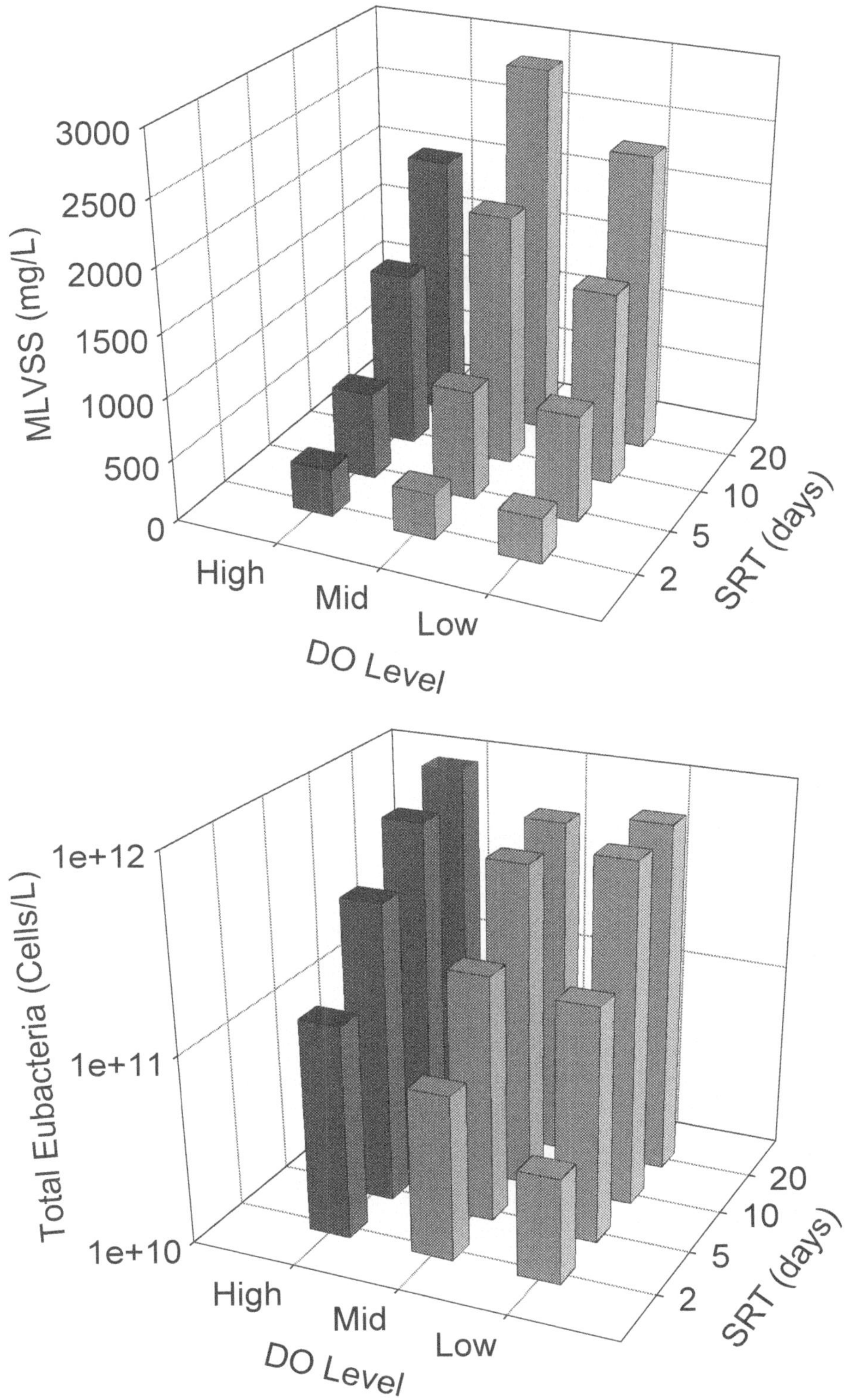

Figure 4-23. Comparison of Trends in MLVSS Concentration and Total Eubacterial Concentration at Varying SRT and DO Levels at 20°C.

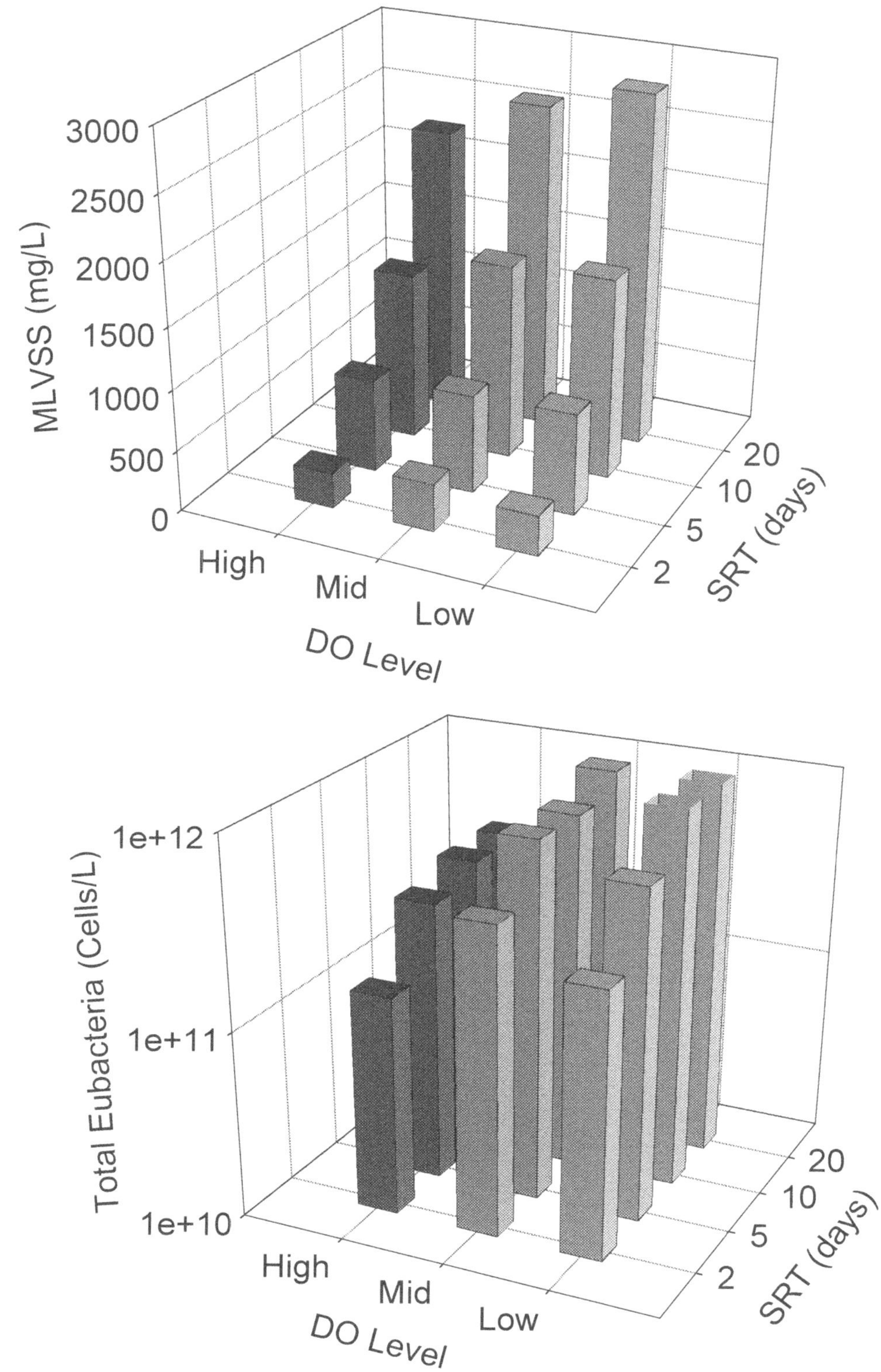

Figure 4-24. Comparison of Trends in MLVSS Concentration and Total Eubacterial Concentration at Varying SRT and DO Levels at 15°C.

interaction between SRT and DO. Under these conditions, both the SRT and DO were found to significantly impact the total eubacterial concentration. The SRT was positively correlated with eubacterial concentration, while the DO level was negatively correlated. SRT was found to have the largest relative effect on the total eubacterial concentration.

It is unclear why the eubacterial concentrations increased with decreasing DO level. One possible explanation is that population shifts occurred in the reactors; with dominant bacteria at the lower DO levels having higher yields than those dominant at the high DO levels. Low DO concentrations favor filamentous organisms (Grady et al., 1999), which is indicative of a population shift. A foaming problem was also noted in the reactors during the intermediate and low DO periods. Excessive foaming is caused primarily by bacteria of the genus *Nocardia* and the species *Microthrix parvicella* (Jenkins et al., 1993). Thus, foaming in the reactors also signifies a possible population shift.

4.1.5.2.3 10°C Operation

Table 4-31 lists the average total eubacterial concentrations for each DO level during operation at 10°C, which are very similar to those obtained at the higher temperatures. In general, the total eubacteria cell density decreased with decreasing SRT.

Total eubacterial concentrations at 10°C appeared to be unaffected by changes in DO concentration. Figure 4-25 provides a graphical representation of this data, as well as the MLVSS data for these operating conditions. The concentration of total eubacteria did not appear to be affected by the changes in influent COD that resulted in decreased MLVSS levels during operation at the low and intermediate DO levels.

Table 4-31. Average (±SD) Total Eubacteria Concentrations (cells/L) During Operation at 10°C.

DO Level	High	Intermediate	Low
20-Day SRT	$18.0 (\pm 0.2) \times 10^{11}$	$5.0 (\pm 0.8) \times 10^{11}$	$16.0 (\pm 0.2) \times 10^{11}$
10-Day SRT	$10.0 (\pm 0.2) \times 10^{11}$	$10.0 (\pm 0.3) \times 10^{11}$	$9.5 (\pm 2.4) \times 10^{11}$
5-Day SRT	$4.8 (\pm 1.5) \times 10^{11}$	$4.7 (\pm 0.5) \times 10^{11}$	$6.9 (\pm 2.3) \times 10^{11}$
2-Day SRT	$3.7 (\pm 0.9) \times 10^{11}$	$3.2 (\pm 0.7) \times 10^{11}$	$3.3 (\pm 0.9) \times 10^{11}$

Linear regression analysis of the 10°C total eubacterial cell density data revealed that only 52.5% of the observed variability in measured values could be explained by changes in the influent COD, SRT, DO, and combined effects of SRT and DO. Only SRT was found to significantly impact the concentration of total eubacteria. In contrast to MLVSS, the total eubacterial cell density was not affected by the large decrease in influent COD concentration.

4.1.5.3 Impact of Test Parameters on Eubacteria Variability

The linear regression model developed for all the total eubacteria data from 20°C, 15°C, and 10°C operating periods described the data fairly well (R=0.734), but showed that only 54% of the observed variability could be explained by changes in the influent COD level, SRT, temperature, DO, and the interactions between these variables. The partial regression coefficients

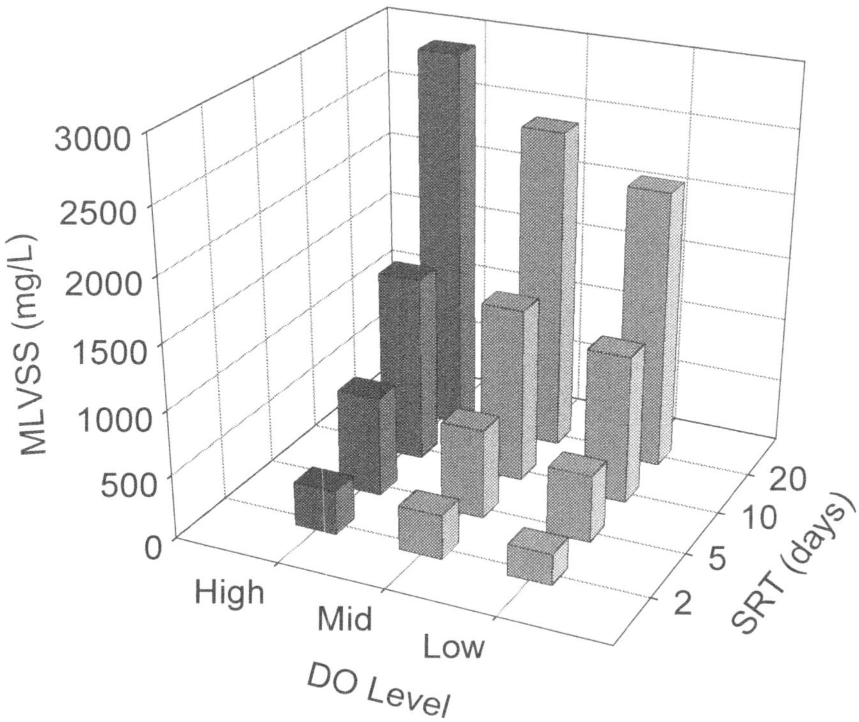

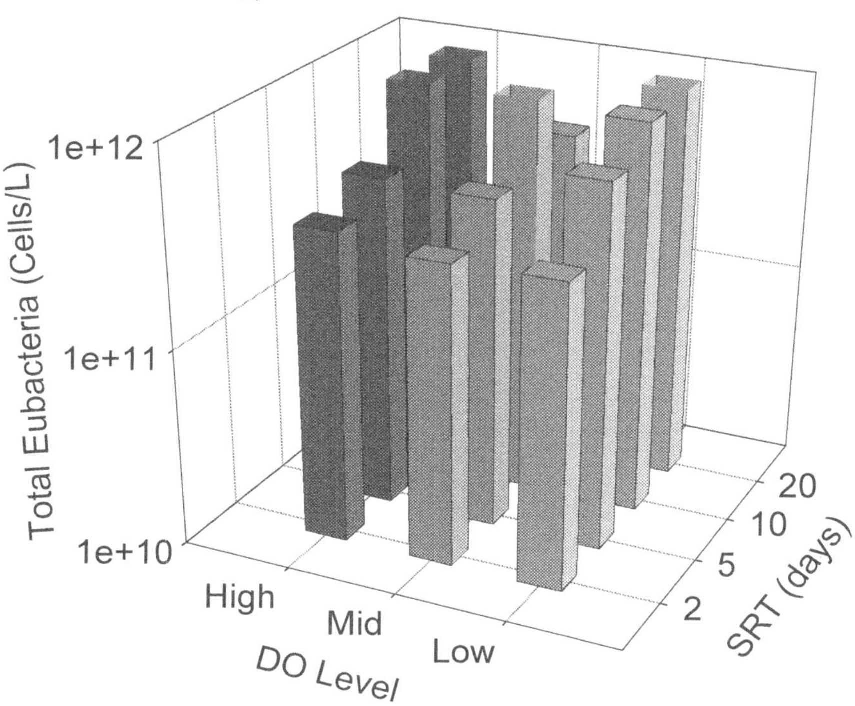

Figure 4-25. Comparison of Trends in MLVSS Concentration and Total Eubacterial Concentration at Varying SRT and DO Levels at 10°C.

WERF

provided unclear results regarding the independent variables yielding the largest influence on total eubacterial concentration. It is believed that this was a result of the strong correlation between the interaction terms and the individual variables, which led to problems of multicollinearity.

At each temperature, the SRT was found to have the most significant impact on concentration of total eubacteria. No clear trends in eubacterial concentration were observed with respect to temperature or DO (Figure 4-26, which includes only steady-state data as identified in Tables 4-26, 4-27, and 4-28). Figure 4-27 shows similar trends for MLVSS. Both total eubacterial and MLVSS concentrations were found to increase with increasing SRT between 2 and 10 days and then level off between 10 and 20 days.

Temperature and DO level were not expected to have a significant impact on the MLVSS or total bacterial concentration in the range of operating conditions tested since heterotrophic populations are not as sensitive as nitrifying bacteria to changes in these conditions. In a study on the effects of temperature on wastewater treatment under aerobic and anoxic conditions, Lishman et al. (2000) found that the change in temperature from 19°C and 14°C did not produce a statistically significant difference in bacterial concentrations within an SBR treatment system.

As shown in Table 4-32, significantly less of the variability in total eubacterial cell numbers (compared to MLVSS numbers) could be explained by the parameters controlled in this study. This indicates that one or more additional factors accounting for a substantial portion of the variability in eubacterial concentration were not included in the linear regression analysis. A potential source of the undescribed variability is the cell weight change in response to variations in operational and environmental parameters. MLVSS is a crude measure of dry cell weight and is maintained at fairly stable levels, through biomass wastage, to maintain a constant SRT regardless of temperature or DO level. In contrast, the real-time PCR assay measures copy numbers of eubacterial 16S rDNA. As discussed previously, in this study it was assumed that each cell contained 3.6 copies of 16S rDNA. However, it has been hypothesized that bacteria with higher maximum specific growth rates have higher ribosomal operon numbers than slow growing bacteria (Condon, 1995; Klappenbach et al., 2000), suggesting that lower SRTs may select for microbial populations with a different number of ribosomal operons. In addition, cell size and weight vary from bacterial species to species and within a species as a result of environmental conditions. Another factor may be that MLVSS is a measure of all filterable organic matter present in the sample and includes protozoa, fungi, etc. While bacteria are responsible for most carbon oxidation, their contribution to MLVSS may vary for different conditions due to predation by protozoa and metazoa (Rittmann and McCarty, 2001). Undescribed variability could also result from the fact that some of the samples included in the analysis were not taken during steady-state operation.

Table 4-32. Summary of Linear Regression Analyses for MLVSS and Total Eubacterial Concentrations.

	MLVSS		Log Total Eubacterial Concentration	
Temperature	R	R^2	R	R^2
All	0.957	0.916	0.734	0.539
20°C	0.926	0.858	0.769	0.591
15°C	0.974	0.949	0.782	0.611
10°C	0.992	0.983	0.724	0.525

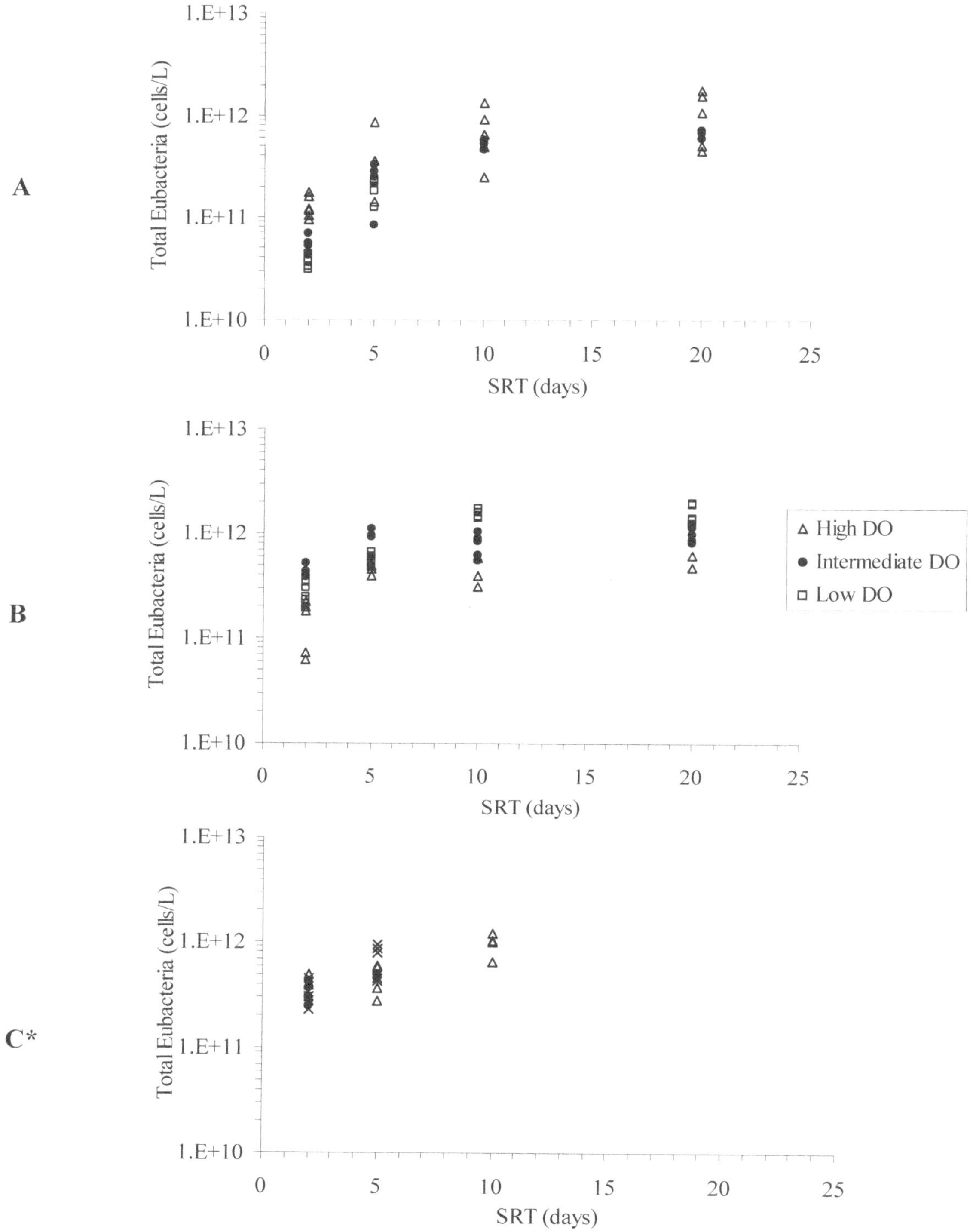

Figure 4-26. Steady-state Concentration of Total Eubacteria as a Function of SRT (A) 20°C, (B) 15°C, (C) 10°C (*20-day SRT data at 10°C did not fit criteria for steady-state and therefore is not included in Plot C).

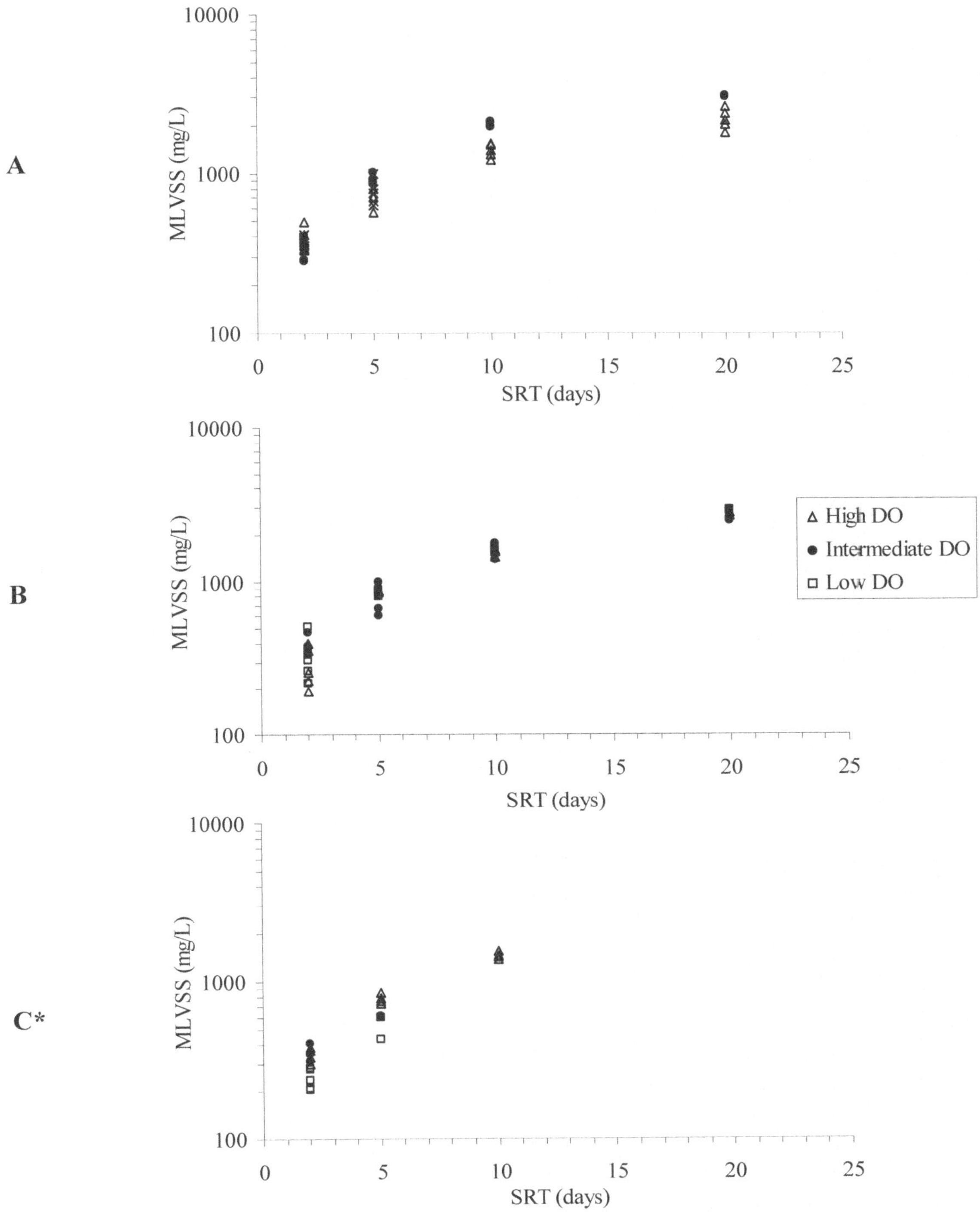

Figure 4-27. Steady-state Concentration of MLVSS as a Function of SRT (A) 20°C, (B) 15°C, (C) 10°C (*20-day SRT data at 10°C did not fit criteria for steady-state and therefore is not included in Plot C).

4.1.5.4 *N. oligotropha*-type *amo*A Gene

Concentrations of *N. oligotropha*-type AOB were calculated from real-time PCR data assuming that *N. oligotropha*-type AOB have two copies of the *amo*A gene per cell (McTavish et al., 1993).

4.1.5.4.1 20°C Operation

Table 4-33 lists the average *N. oligotropha*-type AOB concentrations at each SRT and DO level for the 20°C study, which differ somewhat from those reported in the literature. Harms et al. (2003) and Dionisi et al. (2002a) measured the cell density of *N. oligotropha*-type AOB in 12 monthly mixed liquor samples (SRT=10 days) taken from the full-scale municipal WWTP used to collect influent waste for a bench-scale study. The average concentration of AOB in the WWTP (using the same real-time PCR assay as that utilized in the bench-scale study) was 7.5 (±6.0) x 10^9 cells/L (Harms et al., 2003). Dionisi et al. (2002a), using a cPCR assay, reported lower AOB values (4.3 [±0.6] x 10^7 to 4.1 [±1.3] x 10^8 *N. oligotropha*-type AOB cells/L) for the full-scale municipal WWTP. Daims et al. (2001) measured an average AOB cell density of 9.8 (±1.9) x 10^{10} cells/L in a separate stage nitrifying municipal WWTP using FISH. Lee et al. (2002) reported 1.18 x 10^{11} AOB cells/L measured by performing colony counts on samples taken from a bench-scale activated sludge reactor operated at 20°C, excess DO, an SRT of 10 days, and a C:N ratio of 12.5, which is comparable to the average value for our study. Wagner et al. (1996) observed a similar average ammonia-oxidizer concentration of 6.6 (±0.3) x 10^{11} cells/L mixed liquor based on FISH measurements. However, this value was obtained at an industrial WWTP that receives sewage with ammonia concentrations two orders of magnitude higher than those measured in this study.

Table 4-33. Average (±SD) *N. oligotropha*-type *amo*A Gene Concentrations (cells/L) During Operation at 20°C.

DO Level	High	Intermediate	Low
20-Day SRT	39.0 (±4.6) x 10^8	13.0 (±2.1) x 10^8	24.0 (±2.0) x 10^8
10-Day SRT	18.0 (±1.1) x 10^8	5.0 (±1.1) x 10^8	7.4 (±6.7) x 10^8
5-Day SRT	2.5 (±2.3) x 10^8	2.5 (±3.0) x 10^8	5.7 (±8.1) x 10^8
2-Day SRT	7.6 (±3.9) x 10^8	3.7 (±2.9) x 10^8	1.4 (±0.3) x 10^8

The DO level did not appear to have a significant effect on the *N. oligotropha*-type AOB population, with the exception of the 2-day SRT reactor. Multiple linear regression analyses performed on the data from each SRT individually showed that DO had a statistically significant impact on AOB cell numbers in the 2-day reactor, but not in the 20-, 10-, and 5-day SRT reactors at the 95% confidence interval. Figure 4-28 shows a comparison of the trends in ammonia oxidation rate (for the dates when molecular sampling was performed) and *N. oligotropha*-type AOB concentration as a function of SRT and DO at 20°C. As shown in this figure, the rate of ammonia oxidation slows in all four SRT reactors at the lowest DO level; however, this trend was not reflected in the *N. oligotropha*-type AOB concentration in the 20-, 10-, and 5-day SRT reactors. The trends in *N. oligotropha*-type AOB concentration do mirror the trends in ammonia oxidation rate in the 2-day SRT reactor, which shows a significant drop in nitrification performance at each DO level.

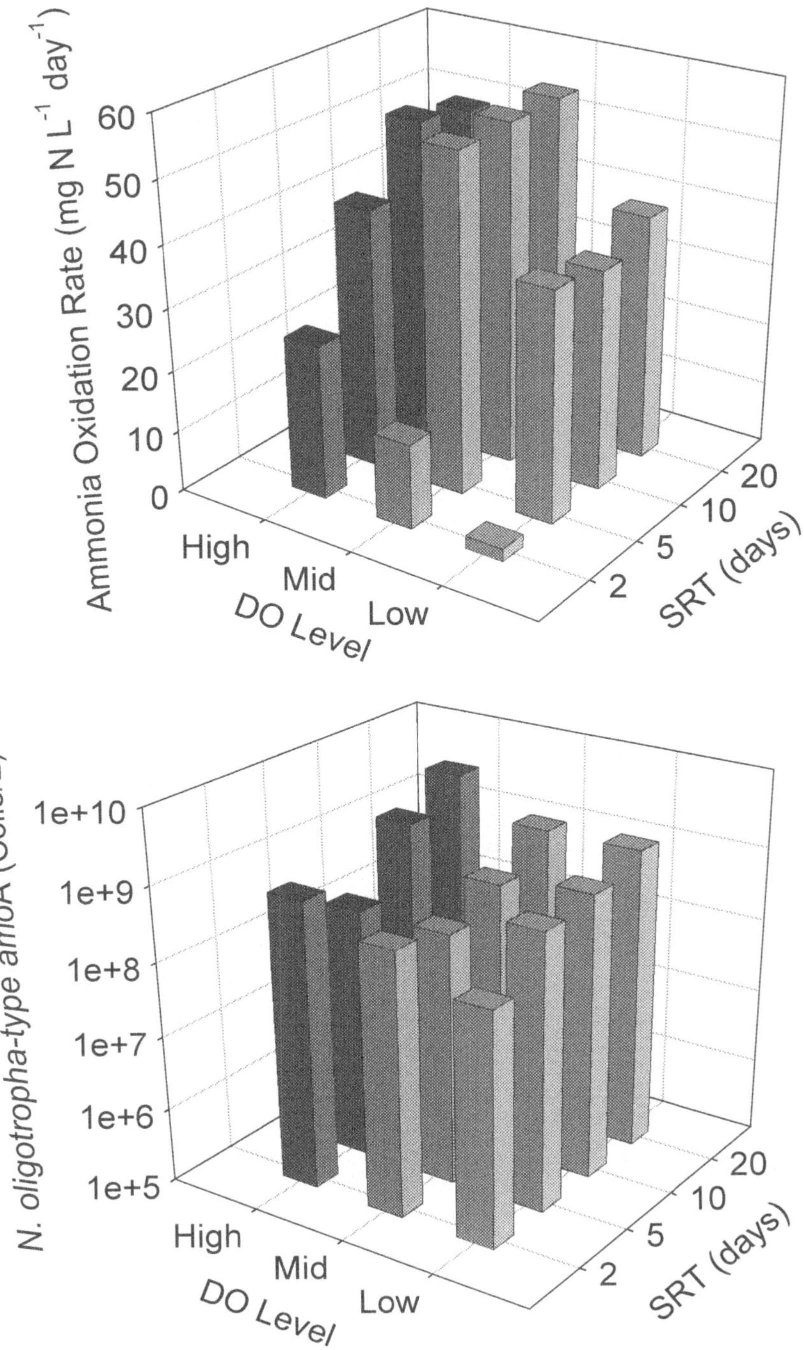

Figure 4-28. Comparison of Trends in Ammonia Oxidation Rate and *N. oligotropha*-type AOB Concentration at Varying SRT and DO Levels at 20°C.

The multiple linear regression model developed for the 20°C *N. oligotropha*-type AOB concentrations did not describe the data very well (R=0.59) and showed that only 35% of the variability in *N. oligotropha*-type AOB cell density could be attributed to changes in the SRT, DO, influent ammonia concentration, and the interaction between SRT and DO. SRT, DO, and influent ammonia concentration were found to significantly affect ammonia oxidation rates at the 95% confidence level. The magnitude of the standardized partial regression coefficients revealed that SRT had the largest relative effect on the AOB concentration, followed by the DO, and then influent ammonia concentration.

4.1.5.4.2 15°C Operation

Average *N. oligotropha*-type AOB concentrations are shown in Table 4-34 for the 15°C study. Comparison of the data in Tables 4-33 and 4-34 revealed that *N. oligotropha*-type AOB concentrations decreased when the temperature was lowered from 20°C to 15°C. This was true for all SRTs and DO levels with the exception of the 5-day SRT reactor at the high and intermediate DO levels. The other SRT reactors showed a one- to 10-fold decrease in cell numbers. Multiple linear regression analyses performed on data taken during operation at 20°C and 15°C for each SRT individually showed that the effect of the change in temperature significantly affected the *N. oligotropha*-type AOB concentrations at the 95% confidence interval for the 20-, 10-, and 2-day SRT reactors, but did not significantly affect the 5-day SRT reactor. During operation at the high and intermediate DO levels at 15°C, no trend was observed in the *N. oligotropha*-type AOB concentration with respect to SRT. At the low DO level, SRT was found to significantly impact the cell density of AOB at the 95% confidence interval based on multiple linear regression analysis. Average *N. oligotropha*-type AOB concentration did decrease with decreasing DO level, except in the 10-day SRT reactor at the low DO level. Figure 4-29 reveals that the trends in ammonia oxidation rates were only mirrored by the *N. oligotropha*-type AOB concentrations in the 5- and 2-day SRT reactors.

Table 4-34. Average (±SD) *N. oligotropha*-type *amo*A Gene Concentrations (cells/L) During Operation at 15°C.

DO Level	High	Intermediate	Low
20-Day SRT	6.9 (±3.7) x 10^8	4.7 (±1.7) x 10^8	4.1 (±1.7) x 10^8
10-Day SRT	4.1 (±2.2) x 10^8	1.8 (±0.9) x 10^8	4.6 (±2.0) x 10^8
5-Day SRT	6.9 (±6.2) x 10^8	6.2 (±2.6) x 10^8	2.8 (±1.2) x 10^8
2-Day SRT	4.1 (±4.0) x 10^8	3.2 (±1.8) x 10^8	0.14 (±0.8) x 10^8

When the 15°C *N. oligotropha*-type AOB data was described by a multiple linear regression model, only 23% of the observed variability in measured AOB concentrations could be explained by changes in the SRT, DO, influent ammonia concentration, and the combined effects of SRT and DO. SRT and DO contributed to the model in a statistically significant manner. Again SRT was found to have a larger relative effect on the AOB concentration than the DO.

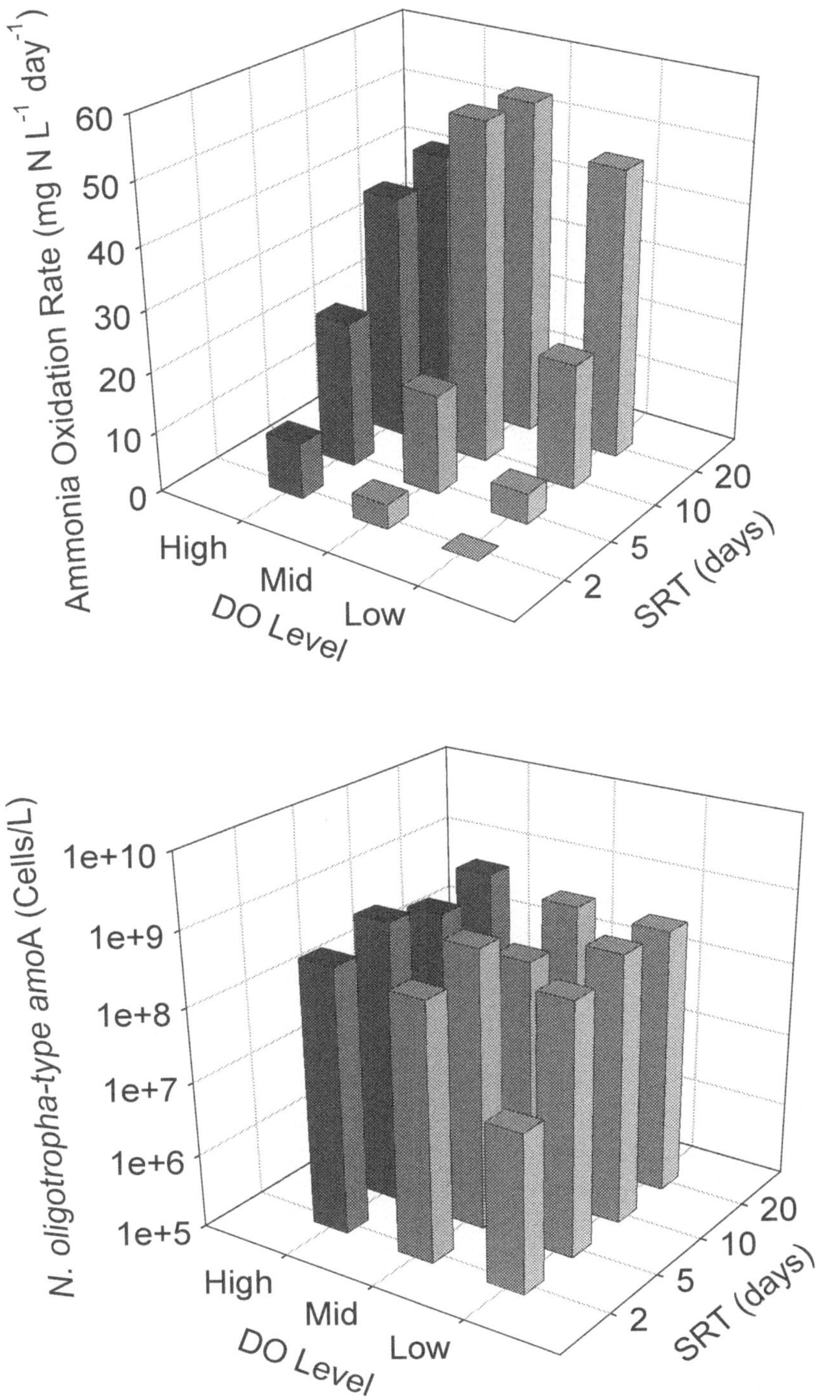

Figure 4-29. Comparison of Trends in Ammonia Oxidation Rate and *N. oligotropha*-type AOB Concentration at Varying SRT and DO Levels at 15°C.

4.1.5.4.3 10°C Operation

The average *N. oligotropha*-type AOB are listed in Table 4-35 for the 10°C study. Tables 4-34 and 4-35 show again that as temperature decreased the concentration of *N. oligotropha*-type AOB also decreased. This can be seen in all SRT reactors at all DO levels with the exception of the 10-day SRT reactor at the intermediate and low DO levels. Multiple linear regression analyses performed on data obtained during operation at 15°C and 10°C data for each SRT individually showed that the temperature drop significantly affected AOB cell densities in the 20-, 5-, and 2-day SRT reactors at the 95% confidence level, but not for the 10-day SRT reactor. In general, the *N. oligotropha*-type AOB concentration decreased with decreasing SRT during 10°C operation. In the 20-day SRT, the concentration of *N. oligotropha*-type AOB decreased with decreasing DO. The opposite trend was observed in the 10-, 5-, and 2-day SRT reactors. Figure 4-30 shows these trends. This plot shows that a slight decrease in ammonia oxidation rate occurred in the 20-day SRT reactor, which corresponded to the decrease in *N. oligotropha*-type AOB level in this reactor. Interestingly, ammonia oxidation rates in the 10-day SRT reactor actually increased when the DO level was lowered. The *N. oligotropha*-type AOB levels obtained from real-time PCR reflect this trend. The 2-day SRT reactor was at complete nitrification failure throughout the 10°C study. This was reflected in a fairly stable *N. oligotropha*-type AOB concentration of approximately 2 x 10^6 cells/L, which was over two orders of magnitude lower than the concentration observed when nitrification was occurring in this reactor. Figure 4-30 also shows that ammonia oxidation rates followed the same general trend as the *N. oligotropha*-type AOB concentration, with the rate decreasing as SRT decreased and no consistent correlation between oxidation rates and DO level.

Table 4-35. Average (±SD) *N. oligotropha*-type *amo*A Gene Concentrations (cells/L) During Operation at 10°C.

DO Level	High	Intermediate	Low
20-Day SRT	4.1 (±1.8) x 10^8	1.2 (±2.2) x 10^8	0.78 (±7.9) x 10^8
10-Day SRT	0.91 (±5.2) x 10^8	1.9 (±1.5) x 10^8	25.0 (±0.9) x 10^8
5-Day SRT	0.39 (±2.3) x 10^8	0.78 (±2.9) x 10^8	1.8 (±0.8) x 10^8
2-Day SRT	0.013 (±1.2) x 10^8	0.023 (±2.7) x 10^8	0.025 (±2.0) x 10^8

According to multiple linear regression analysis performed on the measured *N. oligotropha*-type AOB concentrations during 10°C operation, only 24% of the variability observed in *N. oligotropha*-type AOB concentration could be attributed to changes in the SRT, DO, and influent ammonia concentration. Of the controlled variables, only the SRT was found to significantly affect the *N. oligotropha*-type AOB cell density at the 95% confidence level.

4.1.5.5 Impact of Test Parameters on *N. oligotropha*-type AOB Variability

When data from all operational conditions were included in the linear regression model, only 18% of the observed variability in *N. oligotropha*-type AOB concentration could be explained by changes in the SRT, temperature, DO, and influent ammonia. Only SRT and temperature significantly affected the concentration of *N. oligotropha*-type AOB, with little difference between the relative effects of the two parameters.

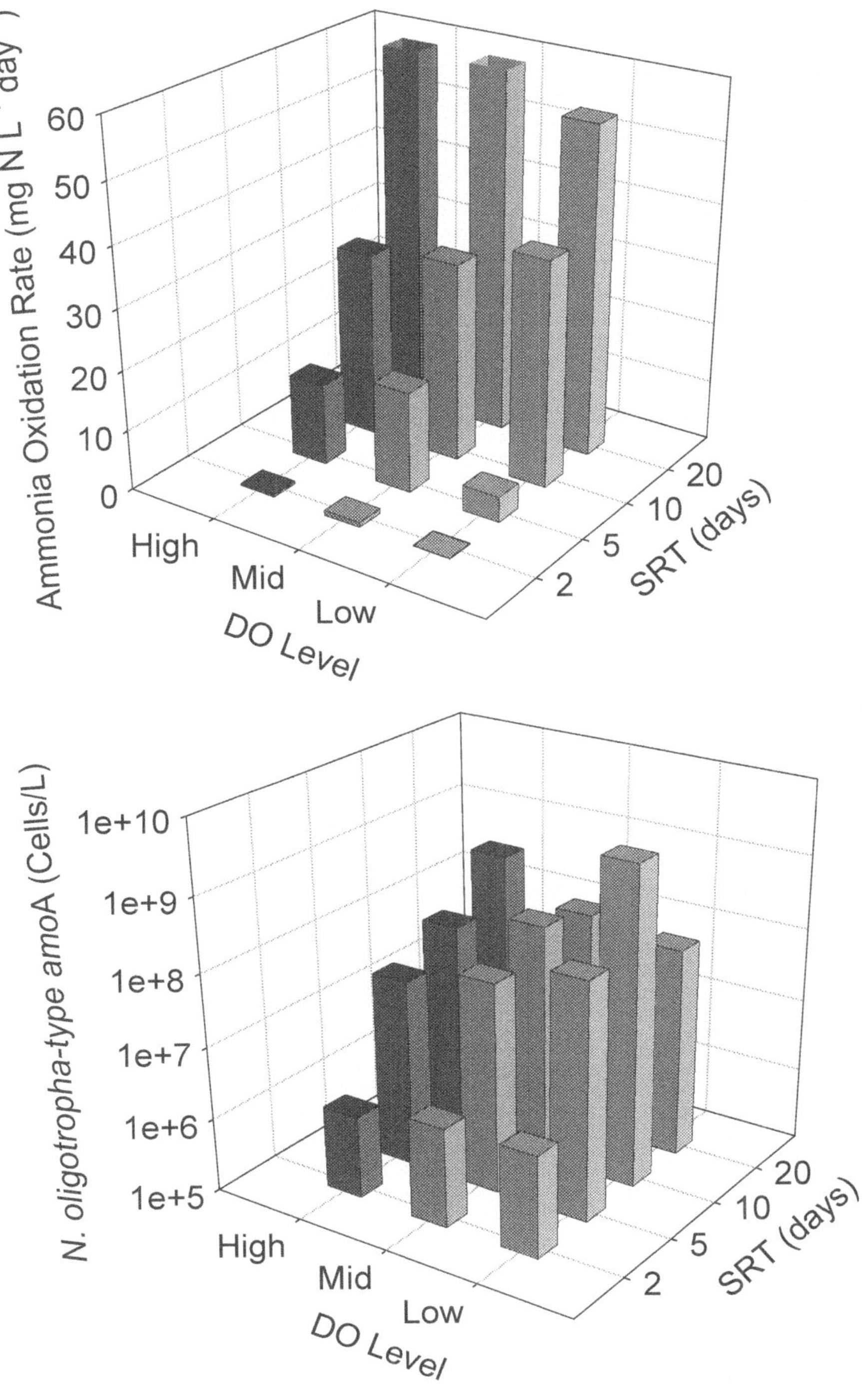

Figure 4-30. Comparison of Trends in Ammonia Oxidation Rate and *N. oligotropha*-type AOB Concentration at Varying SRT and DO Levels at 10°C.

Table 4-36 lists the correlation coefficients and R^2 values obtained from the linear regression analyses performed on the ammonia oxidation rate and *N. oligotropha*-type AOB data when all variables and compound effects of variables were included in the analyses. Significantly less of the variability in *N. oligotropha*-type AOB concentrations (relative to ammonia oxidation rates) could be explained by the parameters controlled in this study. A likely explanation for this is that the real-time PCR assay only measured a subset of the population, and population shifts could occur due to changes in operating and environmental conditions. Therefore, organisms that were not being measured could have significantly contributed to the removal of substrate.

Table 4-36. Regression Analyses for Ammonia Oxidation Rate and *N. oligotropha*-type AOB Concentrations.

	Ammonia Oxidation Rate		Log *N. oligotropha*-type AOB	
Temperature	R	R^2	R	R^2
All	0.850	0.722	0.678	0.460
20°C	0.563	0.317	0.590	0.349
15°C	0.854	0.730	0.480	0.230
10°C	0.981	0.962	0.533	0.283

4.1.5.6 Comparison of Data from Ammonia Oxidation and AOB Concentrations

Attempts were made to compare the percentage of AOB measured in the reactors using real-time PCR for the *N. oligotropha*-type *amo*A gene and theoretical values estimated from design equations using typical kinetic parameters. The percentages of AOB measured in the reactors using real-time PCR were computed by dividing the *N. oligotropha*-type AOB concentration (cells/L) by the total eubacterial 16S concentration (cells/L) for the steady-state samples noted in Tables 4-26, 4-27, and 4-28. The theoretical values were calculated using the approach outlined in Rittmann and McCarty (2001). For the theoretical approach, the volatile suspended solids (VSS) production rates for all types of biomass (heterotrophs and ammonia oxidizers) were computed according the following equation.

$$\frac{\Delta X_v}{\Delta t} = Q\left(S^o - S\right)Y\frac{1+\left(1-f_d\right)b\theta_x}{1+b\theta_x} \qquad (4.1.5.6\text{-}1)$$

where:

$\frac{\Delta X_v}{\Delta t}$ = VSS production rate (mg VSS/day)
Q = hydraulic flow rate (L/day)
S^o = influent substrate concentration (mg/L)
S = effluent substrate concentration (mg/L)
Y = true yield (mg/mg)
f_d = biodegradable fraction (unitless)
b = endogenous decay coefficient (days^{-1})
θ_x = SRT (days)

The percentage of AOB was then estimated by taking the ratio of the VSS production rate for ammonia oxidizers to the total biomass produced. Average influent and effluent parameters measured during the steady-state operating periods in which molecular sampling occurred (Tables 4-37 and 4-38) were used in theoretical calculations. A value of 0.8 was used for the biodegradable fraction (f_d) of each microbial population. BOD values were estimated from COD values according to the following equation (Grady et al. 1999):

$$COD_T = 2.1(BOD_5) \quad (4.1.5.6\text{-}2)$$

where:

COD_T = total COD (mg/L)
BOD_5 = 5-day BOD (mg/L)

Typical kinetic parameters used in the calculations were taken from Rittmann and McCarty (2001) and are summarized in Tables 4-39 and 4-40 for heterotrophs and ammonia oxidizers, respectively, as a function of temperature. Table 4-41 provides a comparison of the percentage of AOB measured in the reactors using real-time PCR for the *N. oligotropha*-type *amo*A gene and theoretical values estimated from design equations using typical kinetic parameters. The values measured using PCR are much lower than the theoretically predicted values. The extremely low percentage of ammonia oxidizers suggests that AOB other than *N. oligotropha*-type were dominant in the bench-scale system. *N. oligotropha* was the only AOB identified in mixed liquor samples from the full-scale municipal WWTP and bench-scale reactors at the beginning of this study. However, since multiple AOB exist (at least 15 strains) (Purkhold et al., 2000), additional work is necessary to identify all of the major AOB species present in the reactors.

In addition to the measured percentages of AOB in this study being much lower than the calculated values, they are also significantly lower than those reported in the literature. You et al. (2003) reported that *Nitrosospira*, which was identified as the sole ammonia-oxidizer in an A20 pilot plant BNR system, accounted for 4.1% of the total biomass measured using a cloning-denaturing gradient gel electrophoresis (DGGE) method. The A20 system, which contains anaerobic, anoxic, and aerobic phases to carry out carbon, nitrogen, and phosphorous removal, was operated at 20°C and a 10-day SRT. This system was used to treat synthetic wastewater with an influent concentration COD of 300 mg/L and an influent total nitrogen

Table 4-37. Average Steady-State Influent Parameters During High DO Operation.

Temperature	Influent parameter	SRT (days) 20	10	5	2
20°C	COD (mg COD/L)	215	220	185	236
	BOD (mg BOD/L)	102	105	88	112
	NH^+_4-N (mg NH^+_4-N/L)	19	21	19	21
15°C	COD (mg COD/L)	141	184	184	191
	BOD (mg BOD/L)	67	88	88	91
	NH^+_4-N (mg NH^+_4-N/L)	16	18	18	16
10°C	COD (mg COD/L)	No steady-state data	174	174	174
	BOD (mg BOD/L)	No steady-state data	83	83	83
	NH^+_4-N (mg NH^+_4-N/L)	No steady-state data	27	27	27

Table 4-38. Average Steady-State Effluent Parameters During High DO Operation.

Temperature	Effluent parameter	SRT (days) 20	10	5	2
20°C	COD (mg COD/L)	27	23	29	33
	BOD (mg BOD/L)	13	11	14	16
	NH^+_4-N (mg NH^+_4-N/L)	<0.1	<0.1	<0.1	8.3
	NO^-_2-N (mg NH^+_4-N/L)	<0.1	<0.1	<0.1	1.3
15°C	COD (mg COD/L)	26	30	33	38
	BOD (mg BOD/L)	12	14	16	18
	NH^+_4-N (mg NH^+_4-N/L)	<0.1	<0.1	5	11
	NO^-_2-N (mg NH^+_4-N/L)	0.21	0.34	0.35	0.39
10°C	COD (mg COD/L)	No steady-state data	35	41	43
	BOD (mg BOD/L)	No steady-state data	17	20	20
	NH^+_4-N (mg NH^+_4-N/L)	No steady-state data	15	20	24
	NO^-_2-N (mg NH^+_4-N/L)	No steady-state data	0.4	0.5	<0.1

Table 4-39. Typical Kinetic Parameters for Heterotrophs by Temperature (Adapted from Rittmann and McCarty, 2001).

Kinetic parameter	20°C	15°C	10°C
Y (mg VSS/mg BOD)	0.45	0.45	0.45
$\hat{q}$ (mg BOD/mg VSS-d)	20	14.3	10.2
K (mg BOD/L)	10	10	10
b ($days^{-1}$)	0.15	0.11	0.076

Table 4-40. Typical Kinetic Parameters for Ammonia-Oxidizers by Temperature (Adapted from Rittmann and McCarty, 2001).

Kinetic parameter	20°C	15°C	10°C
Y (mg VSS/mg NH^+_4-N)	0.33	0.33	0.33
$\hat{q}_n$ (mg NH^+_4-N/mg VSS-d)	2.3	1.7	1.3
K_n (mg NH^+_4-N/L)	1.0	0.57	0.32
b ($days^{-1}$)	0.11	0.082	0.060

Table 4-41. Comparison of % AOB Based on Theoretical Calculations and Real-Time PCR Analysis.

	20°C		15°C		10°C	
SRT (days)	Theoretical	Measured (±SD)	Theoretical	Measured (±SD)	Theoretical	Measured (±SD)
20	13.0%	0.3 (±0.4)%	9.7%	0.12 (±0.02)%	13.3%	0.02 (±0.01)%
10	9.8%	0.3 (±0.2)%	9.1%	0.1 (±0.04)%	10.3%	0.01 (±0.01)%
5	8.9%	0.07 (±0.09)%	8.3%	0.3 (±0.04)%	5.3%	0.01 (±0.001)%
2	6.2%	0.7 (±0.3)%	Unstable Nitrification	0.3 (±0.2)%	Unstable Nitrification	0.0004(±0.0004)%

concentration of 40 mg/L. Based on FISH measurements, the researchers reported an AOB percentage of 5.2 (±3.8)%. McClintock et al. (1993) calculated 3.4% to 7.5% AOB in a pilot-scale activated sludge system that was treating municipal wastewater and operated at SRTs of 1.5 to 15 days on the basis of kinetic values found in the literature.

It is not surprising that a strong correlation between the measured values of ammonia oxidation rate and *N. oligotropha*-type AOB cells does not exist (Figure 4-31), since the *N. oligotropha*-type organisms measured in the bench-scale reactors were not the dominant AOB present. As discussed in the preceding section, there were only rare instances when the same trends were observed in both the ammonia oxidation rate and *N. oligotropha*-type AOB cell density. Table 4-42 lists the correlation coefficients obtained between the logarithm of the *N. oligotropha*-type AOB concentrations and the ammonia oxidation rates for each temperature and DO level. While there were a few incidences when the correlation coefficient exceeded 0.5, such as during operation at the high DO level at 10°C, the *N. oligotropha*-type AOB concentration and ammonia oxidation rate were not highly correlated.

Specific ammonia oxidation rates were computed by dividing the measured oxidation rate by the measured concentration of *N. oligotropha*-type AOB and performing appropriate unit conversions for the steady-state samples shown in Tables 4-26, 4-27, and 4-28. These resulting ammonia oxidation rates are shown in Figure 4-32A, B, and C for the high, intermediate, and low DO levels, respectively. The specific ammonia oxidation rates were, in general, much higher than the reported rates of 0.0018 pmol/cell/hour to 0.023 pmol/cell/hour for pure cultures

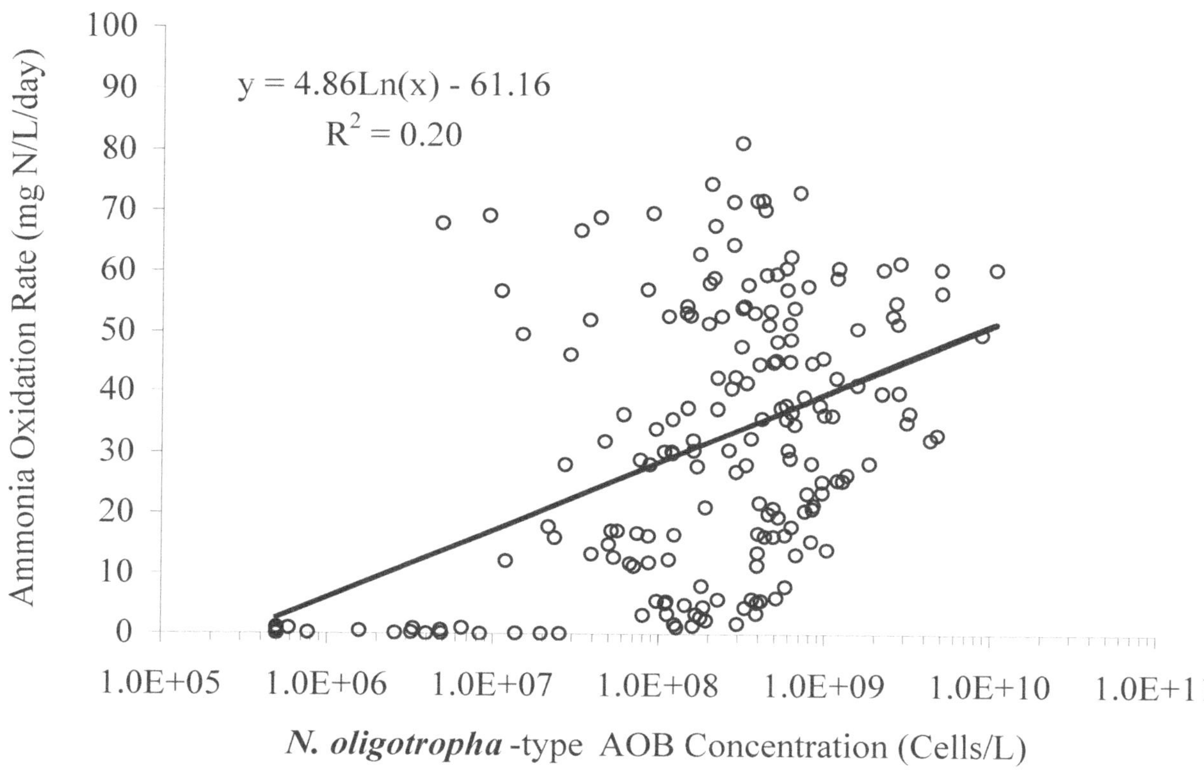

Figure 4-31. Correlation Between Ammonia Oxidation Rate and *N. oligotropha*-type AOB Concentration (Includes Data From All Operational Conditions).

WERF

Table 4-42. Pearson Correlation Coefficients Between *N. oligotropha*-type AOB Concentrations and Ammonia Oxidation Rates.

	DO Level (mg/L)		
Temperature (°C)	High	Intermediate	Low
20	0.32	0.11	0.67*
15	0.42	0.17	0.60
10	0.84	0.41	0.43

*Data only includes samples taken during operation at 0.5 mg/L DO

(Belser, 1979). The high specific ammonia oxidation rates in this study are reflective of the low reported concentrations of AOB. Figure 4-32A and B suggests that the specific ammonia oxidation rate tended to increase with increasing ammonia concentration during operation at the high and intermediate DO levels. Figure 4-32C shows that the specific ammonia oxidation rate did not vary substantially with ammonia concentration and that the low DO rates were much lower than the rates at the higher DO levels. However, due to the unrealistically high specific ammonia oxidation rates measured in this study, it is difficult to draw firm conclusions regarding this information.

4.1.5.7 *Nitrospira* spp. 16S rDNA

Real-time PCR measurements for *Nitrospira spp.* 16S rDNA were made on samples collected from reactors operating at 20°C, 15°C, and 10°C. Since the number of *Nitrospira* 16S rDNA copies per cell is unknown, it was assumed to be based on the average copies of 16S rDNA per cell in other nitrifying bacteria, such as *Nitrosomonas* and *Nitrobacter* (Navarro et al., 1992; Klappenbach et al., 2001).

4.1.5.7.1 20°C Operation

Table 4-43 presents the average *Nitrospira* concentrations in each of the reactors for the 20°C period. In contrast to the measured AOB cell densities, the NOB concentrations listed in this table are very similar to those reported in the literature. Yuan and Blackall (2002) reported *Nitrospira* concentrations of approximately 3 x 10^9 *Nitrospira* cells/L in lab-scale activated sludge reactors treating municipal wastewater using a real-time PCR assay targeting *Nitrospira* 16S rDNA. Hall et al. (2002) applied the same real-time PCR *Nitrospira* assay to samples from five full-scale municipal WWTPs and measured *Nitrospira* cell densities of 2.9 x 10^6, 4.1 x 10^{10}, 1.8 x 10^{12}, 1.2 x 10^7, and 2.3 x 10^{13} cells/L mixed liquor for the WWTPs achieving 48.1%, 98.8%, 99.3%, 99.7%, and 99.8% nitrification, respectively. Ammonia in the influent for all plants ranged from 21 mg/L to 42 mg/L. Harms et al. (2003), using real time PCR, measured an average *Nitrospira* concentration of 3.7 (± 3.5) x 10^{10} cells/L for 12 monthly mixed liquor samples from the full-scale municipal WWTP (SRT=10 days) used to seed the reactors in this bench-scale study. Dionisi et al (2002a) reported *Nitrospira* cell densities of 5.8 (± 0.98) x 10^9 to 5.5 (± 0.2) x 10^{10} cells/L for the same samples using a cPCR assay. Lee et al. (2002) obtained similar results using direct counts of colony forming units in a 10-day SRT bench-scale activated sludge reactor operated at 20°C and excess DO. Those researchers measured 5.44 x 10^{10} NOB cells/L when the reactor was operated at a carbon/nitrogen ratio of 12.5, which is comparable to the ratio observed in this study.

Nitrospira levels remained fairly constant during 20°C operation in the 20- and 10-day SRT reactors at all DO levels and in the 5-day SRT reactor at the high and intermediate DO levels. The average *Nitrospira* concentration in the 5-day SRT reactor decreased by almost one-half at the low DO level. The *Nitrospira* concentration in the 2-day reactor decreased with decreasing DO and was significantly lower than levels measured in the other SRT reactors at this temperature. Multiple linear regression analyses performed on data from each SRT at 20°C confirmed these findings. DO concentration significantly impacted the NOB concentration in the 2- and 5-day SRT reactors, but not in the 20- and 10-day SRT reactors. This relationship can be seen in Figure 4-33, which shows a comparison of the trends in nitrite oxidation efficiency (described previously) and *Nitrospira* concentration as a function of SRT and DO at 20°C. At the high DO level, nitrite oxidation rates were similar in the 20- and 10-day SRT reactors and decreased incrementally as the SRT decreased below 10 days. This trend was mirrored in the *Nitrospira* cell densities. At the intermediate DO level, nearly identical average nitrite oxidation rates were observed in the 20-, 10-, and 5-day SRT reactors. Whereas, at an SRT of 2 days, a much lower average nitrite oxidation rate was observed. This behavior was also reflected in the average *Nitrospira* concentrations. At the low DO level, nitrite oxidation decreased incrementally with SRT, which was also observed in the cell numbers of *Nitrospira* (except at an SRT of 10 days). Multiple linear regression analysis of the low DO level data showed that both SRT and available nitrite concentration exerted a significant impact on the NOB concentrations at the 95% confidence interval. A decline in nitrite-oxidation rate was noted in all SRT reactors at the low DO level, and nitrite-oxidation rates decreased incrementally in the 2-day SRT reactor each time the DO level was lowered. The trends observed in nitrite-oxidation rate, with respect to DO level, in the 5- and 2-day SRT reactors were mirrored by the trends in *Nitrospira* concentration in these reactors. However, the decrease in nitrite oxidation rate at the low DO level in the 20- and 10- day SRT reactors was not reflected in the *Nitrospira* concentration, which may be due to the fact that although a decrease in rate was observed, a decline in nitrite oxidation efficiency did not occur under these conditions (Table 4-19). Multiple linear regression analysis performed on all of the NOB cell densities measured during 20°C operation indicated that both SRT and the available nitrite concentration significantly impacted the concentration of *Nitrospira* at a 95% confidence level, with the available nitrite concentration yielding almost two times more influence than SRT.

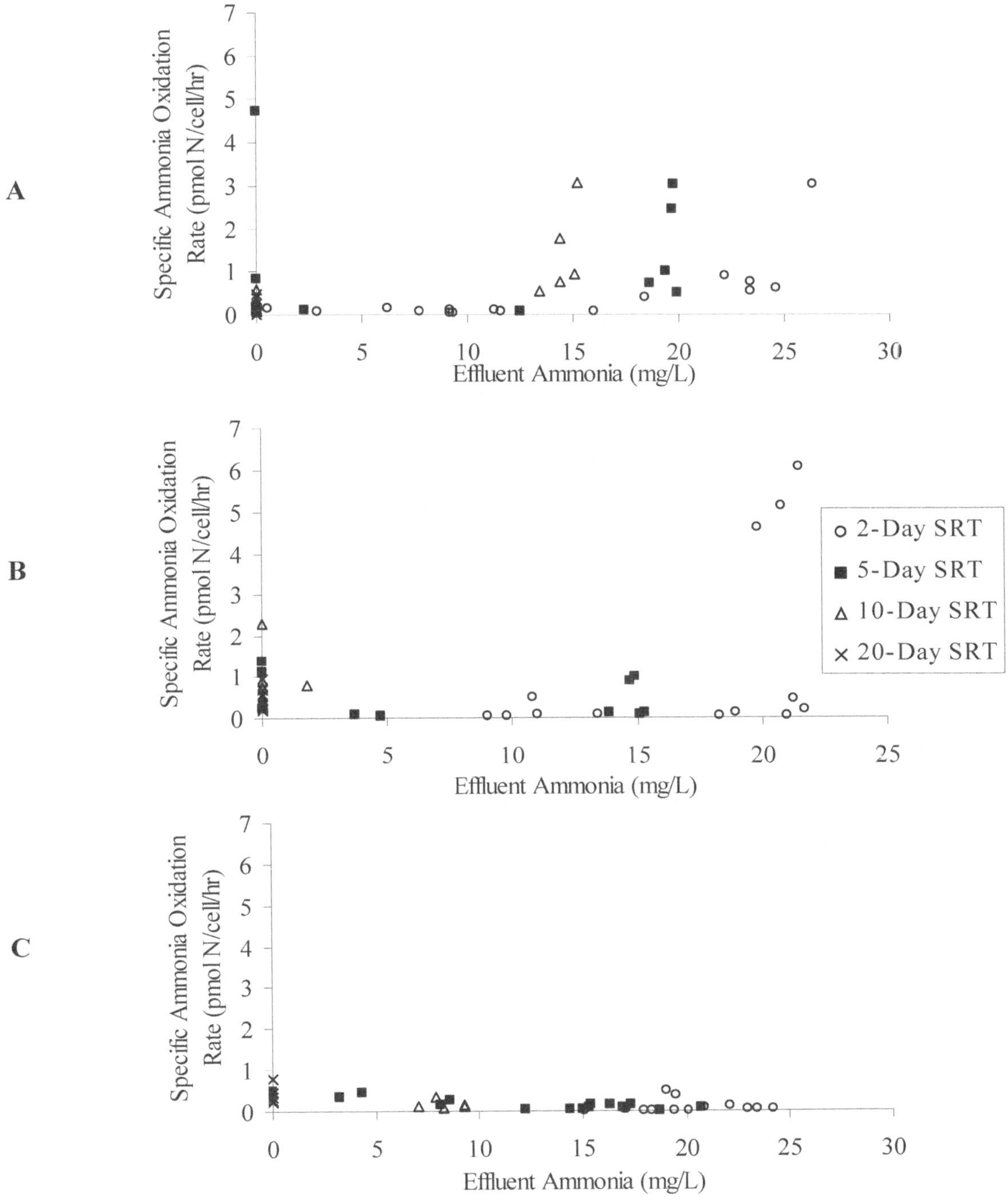

Figure 4-32. Steady-state Specific Ammonia Oxidation Rates as a Function of Effluent Ammonia Concentration (A) High DO, (B) Intermediate DO, (C) Low DO.

Table 4-43. Average (±SD) *Nitrospira* Concentrations (cells/L) During Operation at 20°C.

DO Level	High	Intermediate	Low
20-Day SRT	19.0 (±0.7) x 10^9	10.0 (±0.4) x 10^9	13.0 (±0.7) x 10^9
10-Day SRT	29.0 (±1.0) x 10^9	23.0 (±0.9) x 10^9	25.0 (±1.9) x 10^9
5-Day SRT	13.0 (±0.8) x 10^9	13.0 (±0.4) x 10^9	7.2 (±9.1) x 10^9
2-Day SRT	2.3 (±1.7) x 10^9	0.53 (±7.5) x 10^9	0.19 (±0.2) x 10^9

4.1.5.7.2 15°C Operation

Average *Nitrospira* concentrations in each of the reactors during the 15°C study period are shown in Table 4-44. Comparison of Tables 4-43 and 4-44 show that the *Nitrospira* concentrations in the 20- and 10-day SRT reactors were similar at temperatures of 20°C and 15°C. In contrast, the decrease in temperature yielded a decline in *Nitrospira* levels in the 5-day SRT reactor and regression analysis supported this observation showing that temperature had a significant impact on *Nitrospira* concentration at the 95% confidence interval. The impact of temperature on NOB concentration in the 2-day SRT reactor was inconclusive.

As shown in Table 4-44 and Figure 4-34, average *Nitrospira* concentrations generally decreased with decreasing SRT at 15°C. The *Nitrospira* concentration actually increased in the 20-, 10- and 2-day SRT reactors when the DO levels were lowered. In the 5-day SRT reactor, the *Nitrospira* concentration was relatively unchanged for all DO levels. Multiple linear regression analysis showed that the DO significantly impacted the *Nitrospira* cell densities at a 95% confidence level in the 20-, and 10- day SRT reactors and that the two parameters were negatively correlated. However, in the 5-day SRT reactor only influent available nitrite concentration significantly influenced the *Nitrospira* concentrations. Linear regression results for the 2-day SRT reactor were inconclusive. Inspection of Figure 4-34 shows that the 20-day SRT achieved high rates of nitrite oxidation throughout the 15°C study, while in the 10-day SRT reactor a significant decline in nitrite oxidation rate occurred at the low DO level. The 5- and 2-day SRT reactors showed incremental decreases in nitrite oxidation rate with decreasing DO level; however, these trends were not clearly reflected in *Nitrospira* concentrations.

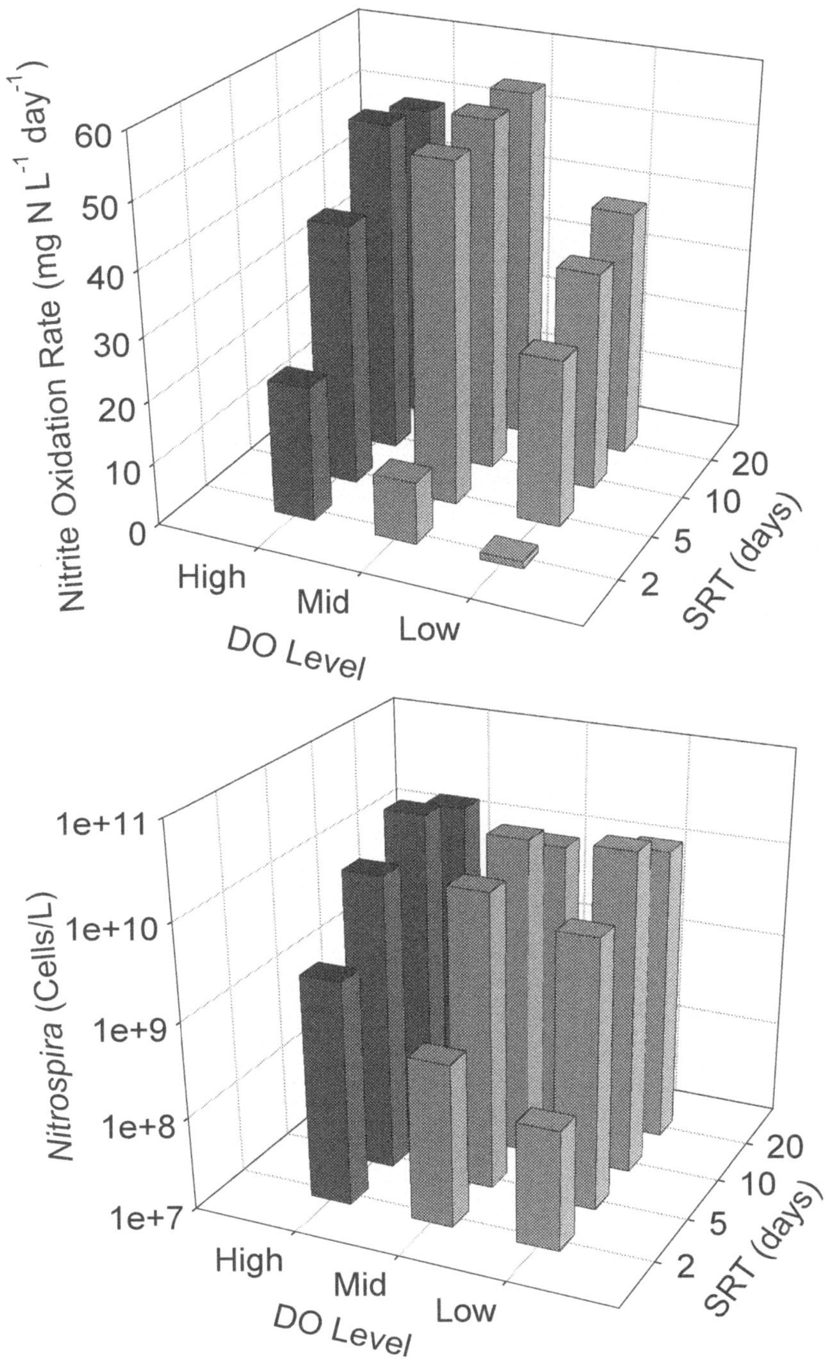

Figure 4-33. Comparison of Trends at 20°C in Nitrite Oxidation Rate and *Nitrospira* Concentration at Varying SRT and DO Levels.

Table 4-44. Average (±SD) *Nitrospira* Concentrations (cells/L) During Operation at 15°C.

DO Level	High	Intermediate	Low
20-Day SRT	12.0 (±1.2) x 10^9	31.0 (±1.0) x 10^9	120.0 (±0.2) x 10^9
10-Day SRT	18.0 (±0.7) x 10^9	26.0 (±0.7) x 10^9	84.0 (±1.4) x 10^9
5-Day SRT	4.3 (±3.2) x 10^9	4.5 (±2.4) x 10^9	2.7 (±0.5) x 10^9
2-Day SRT	0.67 (±3.1) x 10^9	0.81 (±2.1) x 10^9	1.8 (±0.9) x 10^9

When the 15°C *Nitrospira* data was statistically analyzed via regression modeling, 67% of the observed variability in measured NOB concentrations could be explained by changes in SRT, DO, and available nitrite concentration. All three variables were found to significantly impact the *Nitrospira* concentration at the 95% confidence level. DO concentration was found to be negatively correlated with NOB cell density, while SRT and available nitrite concentration were positively correlated and the relative impact of available nitrite was greater than the impacts of DO concentration and SRT by a factor of 1.2.

4.1.5.7.3 10°C Operation

Inspection of the data in Table 4-45 shows that after the temperature was decreased to 10°C, the concentration of *Nitrospira* in 20- and 10-day SRT reactors remained fairly stable at levels similar to those measured at the higher temperatures. This was not true for the 5- and 2-day SRT reactors in which the *Nitrospira* concentration decreased with decreasing temperature, an observation that was supported by the results of regression analysis. Additionally, *Nitrospira* concentration decreased with decreasing SRT (Figure 4-35). There appeared to be no clear correlation between *Nitrospira* concentration and DO level in these reactors when operated at 10°C. The plots in Figure 4-35 show that the 20-day SRT reactor consistently achieved high rates of nitrite oxidation. Again, nitrification performance in the 10-day SRT reactor actually improved when the DO level was lowered. The nitrite oxidation rate in the 5-day SRT reactor decreased significantly at the low DO level. The 2-day SRT reactor was at complete nitrification failure throughout 10°C treatment. The trends in nitrite oxidation rate were reflected in the *Nitrospira* levels found in these reactors with the exception of the

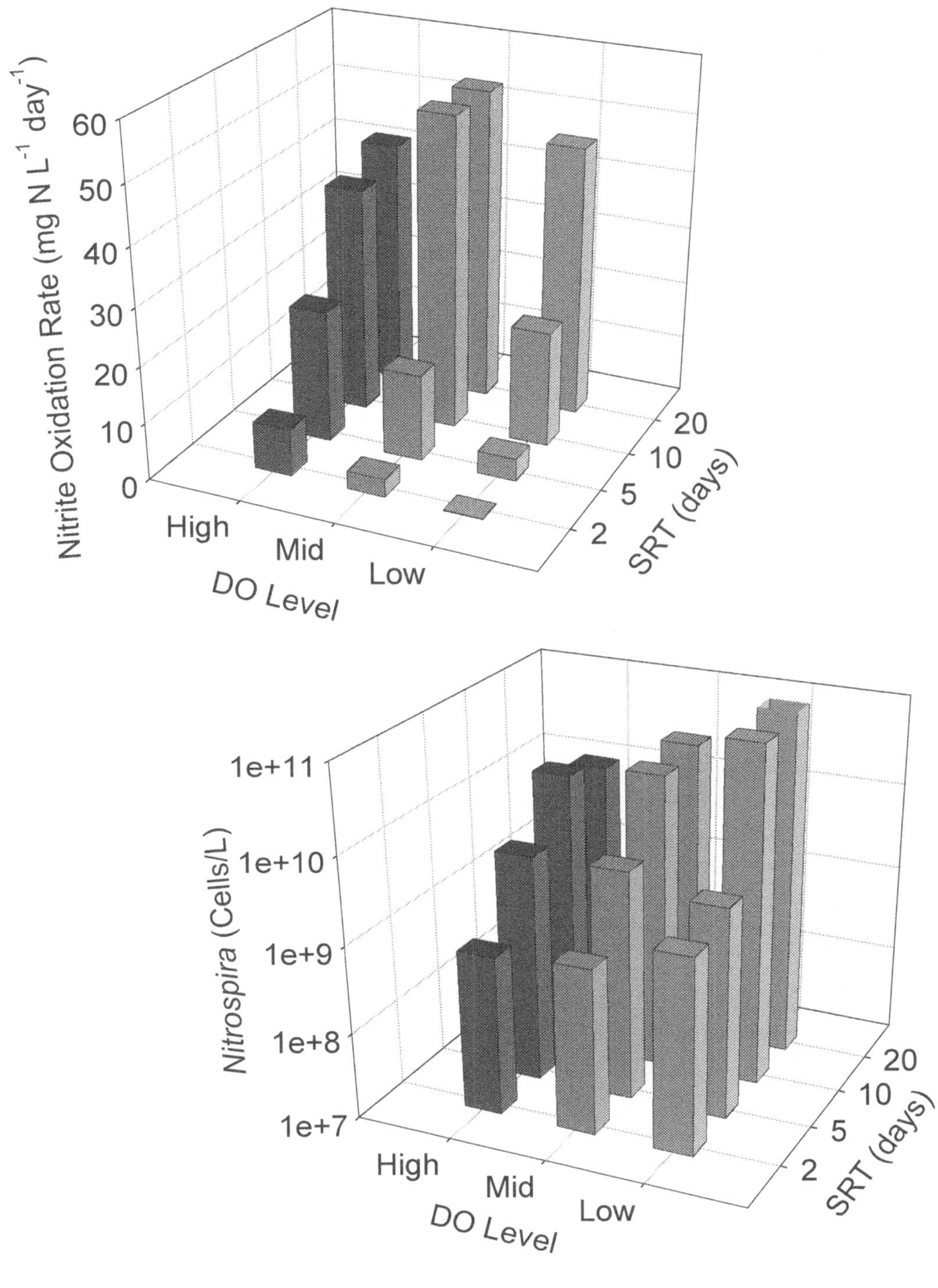

Figure 4-34. Comparison of Trends at 15°C in Nitrite Oxidation Rate and Nitrospira Concentration at Varying SRT and DO Levels.

Table 4-45. Average (±SD) *Nitrospira* Concentrations (cells/L) During Operation at 10°C.

DO Level	High	Intermediate	Low
20-Day SRT	97.0 (±1.9) x 10^9	43.0 (±0.9) x 10^9	53.0 (±0.9) x 10^9
10-Day SRT	20.0 (±0.5) x 10^9	34.0 (±0.3) x 10^9	25.0 (±0.7) x 10^9
5-Day SRT	1.7 (±0.9) x 10^9	0.72 (±2.3) x 10^9	1.2 (±0.4) x 10^9
2-Day SRT	0.41 (±2.3) x 10^9	0.26 (±0.6) x 10^9	0.39 (±1.9) x 10^9

5-day SRT reactor. Multiple linear regression analysis performed on the measured *Nitrospira* concentrations data during 10°C operation indicated that only the available nitrite concentration significantly impacted the NOB cell density at the 95% confidence level.

4.1.5.8 Impact of Test Parameters on Data Variability

When data from all operational conditions were included in the linear regression model, approximately 66% of the observed variability in nitrite oxidation rate could be explained by changes in the SRT, temperature, DO, and available nitrite concentration. All four variables significantly impacted the *Nitrospira* concentration at the 95% confidence interval. The concentration of available nitrite and SRT were positively correlated with *Nitrospira* concentration and yielded similar influence, while temperature and DO were negatively correlated with *Nitrospira* concentration and yielded slightly less influence than available nitrite concentration and SRT.

Table 4-46 lists the correlation coefficients and R^2 values obtained from linear regression analyses performed on the nitrite oxidation rate and *Nitrospira* data when all controlled variables and interaction terms were included in the analysis. While the traditional measure of nitrite-oxidizers was modeled with slightly more accuracy than the molecular measure, the results are somewhat similar. This is a very different finding than that for ammonia oxidation and suggests that the *Nitrospira* assay is representative of the NOB population in the reactors.

4.1.6 Comparison of Nitrite Oxidation and NOB Concentrations

A comparison between the percentages of NOB measured in the reactors using real-time PCR for *Nitrospira* 16S rDNA and theoretical values estimated from design equations using typical kinetic parameters was also performed. The percentages of NOB measured in the reactors using real-time PCR were computed by dividing the *Nitrospira* concentration (cells/L) by the total eubacterial 16S rDNA concentration (cells/L). The theoretical values were calculated using the approach outlined previously for AOB. Steady-state VSS production rates

Table 4-46. Summary of Linear Regression Analyses for Nitrite Oxidation Rate and *Nitrospira* Concentrations.

	Nitrite Oxidation Rate		Log *Nitrospira* Concentration	
Temperature	R	R^2	R	R^2
All	0.971	0.942	0.823	0.677
20°C	0.900	0.810	0.788	0.621
15°C	1.000	0.999	0.834	0.695
10°C	1.000	1.000	0.916	0.840

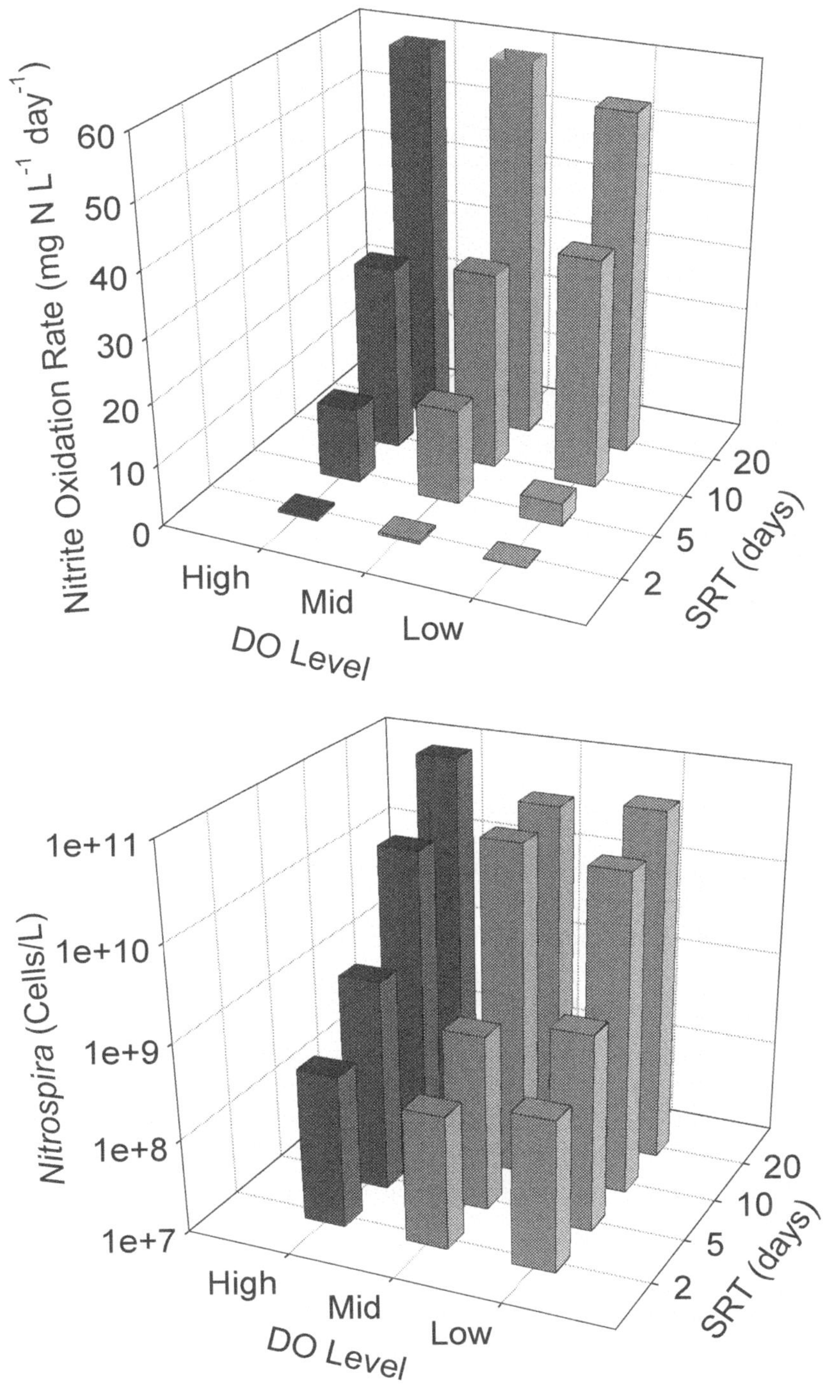

Figure 4-35. Comparison of Trends at 10°C in Nitrite Oxidation Rate and *Nitrospira* Concentration at Varying SRT and DO Levels.

for NOB were estimated using equation 4.1.5.6-1 and the kinetic parameters shown in Table 4-47 (Rittmann and McCarty, 2001). Table 4-48 provides a comparison of NOB measured in the reactors using real-time PCR for *Nitrospira* and theoretical values estimated from design equations using typical kinetic parameters. The values measured using PCR are similar to the theoretically predicted values and are in much better agreement than the AOB values.

Table 4-47. Typical Kinetic Parameters for Nitrite-Oxidizers (Adapted from Rittmann and McCarty, 2001).

Parameter	20°C	15°C	10°C
Y (mg VSS/mg NO_2^--N)	0.083	0.083	0.083
$\hat{q}_n$ (mg NO_2^--N/mg VSS-d)	9.8	7.3	5.5
K_n (mg NO_2^--N/L)	1.3	0.62	0.30
b ($days^{-1}$)	0.11	0.082	0.060

Table 4-48. Comparison of %NOB Based on Theoretical Calculations and Real-Time PCR Analysis.

	20°C		15°C		10°C	
SRT (days)	Theoretical	Measured (±SD)	Theoretical	Measured (±SD)	Theoretical	Measured (±SD)
20	3.2%	2.2 (±1.2)%	4.0%	4.6 (±0.3)%	No steady-state data available	
10	3.2%	5.6 (±3.9)%	2.9%	4.8 (±1.5)%	2.4%	2.1 (±1.1)%
5	3.4%	3.3 (±1.0)%	2.3%	1.4 (±0.4)%	1.1%	0.2 (±0.007)%
2	1.2%	1.5 (±1.0)%	0.4%	0.6 (±0.5)%	Nitrification Failure	0.1 (±0.1)%

The close agreement between measured and calculated values suggests that *Nitrospira spp.* were the dominant nitrite oxidizers in the bench-scale system. The values reported in Table 4-48 are also consistent with those reported in the literature. You et al. (2003), using a cloning-DGGE method, found that *Nitrospira* accounted for 2.1% of the total eubacterial population in a pilot plant A20 BNR system operated at 20°C and an SRT of 10 days. Analysis of the nitrifier abundance in the same treatment system by FISH yielded similar results (2.3 [±0.8]% nitrite-oxidizers in the activated sludge). Rittmann et al. (1999) conducted a study of full-scale WWTPs in France and the Netherlands and measured 4.0% to 8.5% nitrifiers (both AOB and NOB) in nitrifying plants and 0.5% to 0.8% nitrifiers in plants in which no nitrification was measured. These values were determined by slot blot hybridization and FISH.

Figure 4-36 shows nitrite oxidation rate as a function of the *Nitrospira* concentration for all conditions studied. A strong logarithmic relationship exists between nitrite oxidation rate and NOB concentration. The Pearson correlation coefficient (R) for these two parameters was 0.78. A similar plot was obtained by Wong-Chong and Loehr (1975) for *Nitrobacter* concentration and nitrite oxidation rate in batch experiments using enriched acclimated nitrifying activated sludge cultures. Those researchers observed a curve in the form of a Michaelis-Menten saturation plot. Figure 4-36 also indicates a saturation relationship in which the overall nitrite oxidation rate increases with increasing *Nitrospira* levels until a maximum rate is achieved that is independent of NOB cell concentration.

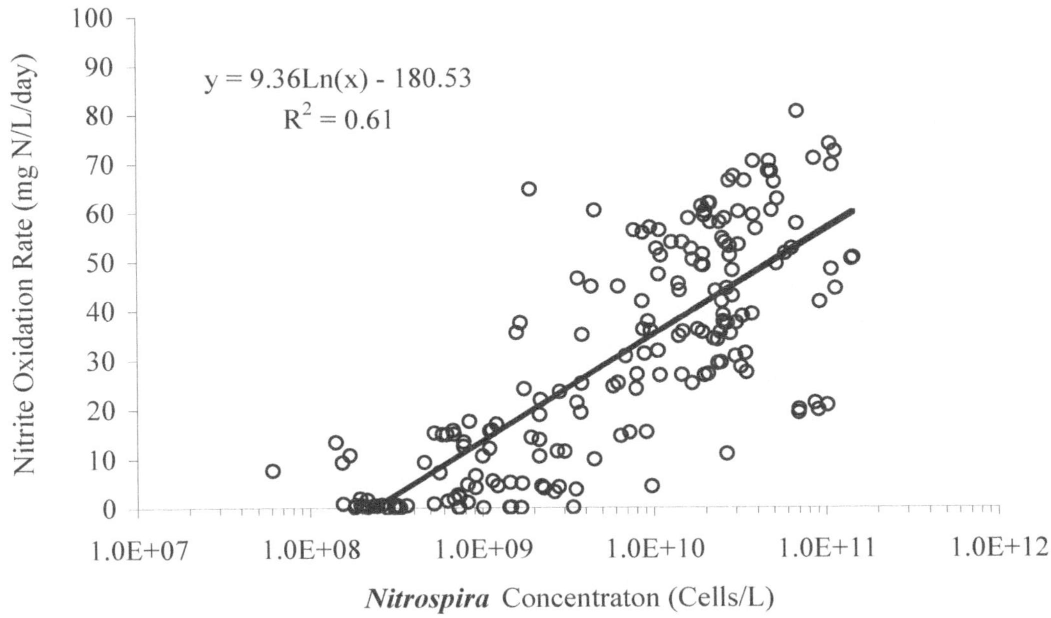

Figure 4-36. Correlation Between Nitrite Oxidation Rate and *Nitrospira* Concentration (Includes Data From All Conditions Studied).

The strength of the correlation between nitrite oxidation rates and *Nitrospira* concentration increased when the data from each temperature and DO level was considered separately. Table 4-49 shows the Pearson correlation coefficient between these two parameters for each set of conditions and shows that *Nitrospira* concentration tracked nitrite oxidation rate quite well with R values greater than 0.86 in all cases. These values are much higher than those obtained for *N. oligotropha*-type AOB concentrations and ammonia-oxidation rates.

Specific nitrite oxidation rates were computed by dividing the measured oxidation rate by the measured concentration of *Nitrospira spp.* NOB and performing appropriate unit conversions for the steady-state samples indicated in Tables 4-26, 4-27, and 4-28. The resulting nitrite oxidation rates are shown in Figure 4-37A, B, and C for the high, intermediate, and low DO levels, respectively. The specific nitrite oxidation rates were in the range of reported rates (0.009 pmol/cell/hour to 0.042 pmol/cell/hour) for pure cultures (Belser, 1979). In general,

Table 4-49. Correlation Coefficients Between *Nitrospira* Concentrations and Nitrite Oxidation Rates.

	DO Level		
Temperature (°C)	High	Intermediate	Low
20	0.86	0.93	0.86*
15	0.87	0.89	0.88
10	0.92	0.87	0.97

*Data only includes samples taken during operation at 0.5 mg/L DO

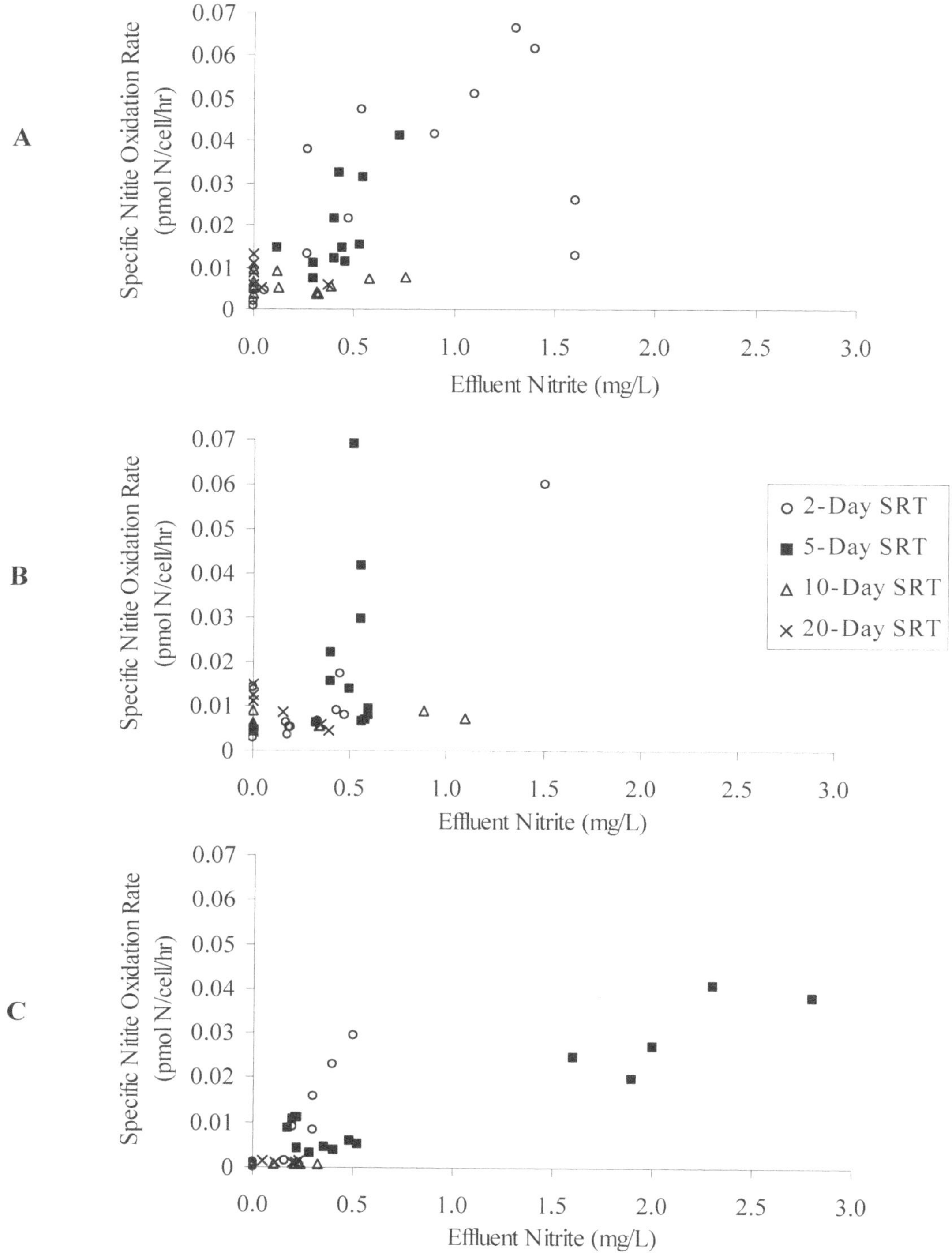

Figure 4-37. Steady-state Specific Nitrite Oxidation Rates as a Function of Effluent Nitrite Concentration (A) High DO, (B) Intermediate DO, (C) Low DO.

Figure 4-37 suggests that the specific nitrite-oxidation rate tended to increase with increasing nitrite concentration for all DO levels. However, Figure 4-37C shows lower specific nitrite oxidation rates were measured at the low DO level compared to the intermediate and high DO levels (Figures 4-37A and B).

Interestingly, *Nitrospira* concentrations were also highly correlated with ammonia oxidation rates (Figure 4-38 and Table 4-50). This is logical since NOB are dependent upon the activity of AOB for the production of their growth substrate (nitrite-N). This suggests that *Nitrospira* cell densities may serve as an indicator for nitrification treatment performance, and the monitoring of AOB may not be necessary. Hall et al. (2002) obtained a similar finding for *Nitrospira* concentrations in samples from five full-scale municipal WWTPs. *Nitrospira* cell densities of 2.9×10^3, 4.1×10^7, 1.8×10^9, 1.2×10^4, and 2.3×10^{10} cells/mL mixed liquor were reported for the WWTPs achieving 48.1%, 98.8%, 99.3%, 99.7%, and 99.8% nitrification, respectively. Hall et al. (2002) noted that, on the surface, nitrification performance appeared to be correlated with *Nitrospira* cell numbers for the plants investigated.

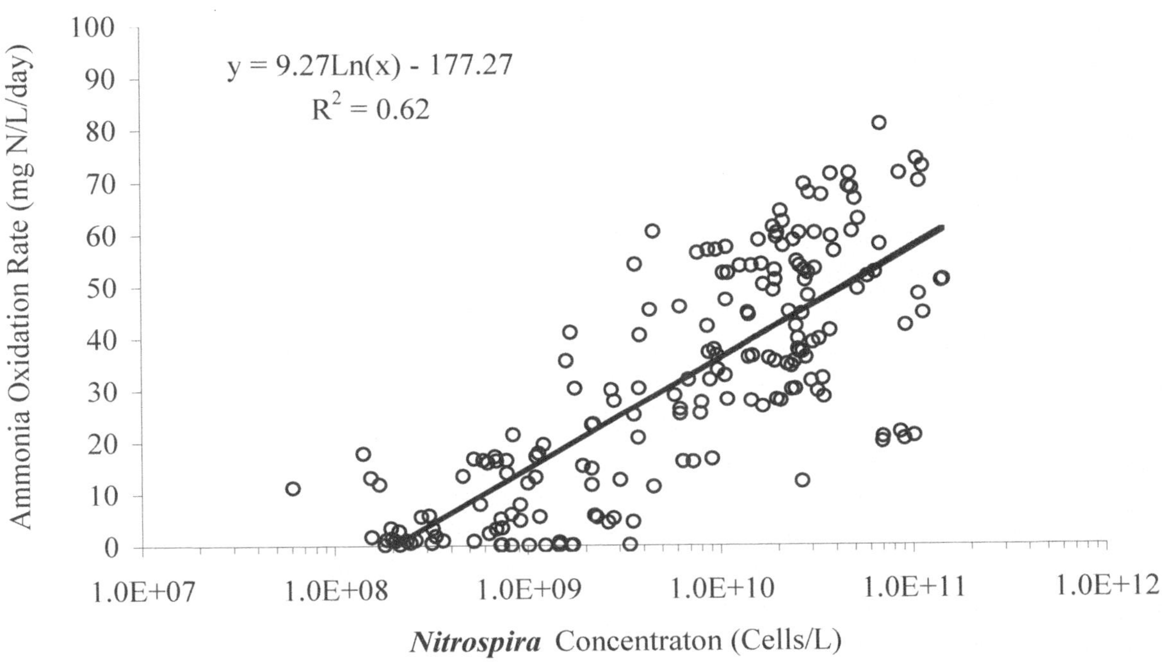

Figure 4-38. Correlation Between Ammonia Oxidation Rate and *Nitrospira* Concentration (Includes Data From All Conditions Studied).

Table 4-50. Correlation Coefficients Between *Nitrospira* Concentrations and Ammonia Oxidation Rates.

	DO Level		
Temperature (°C)	High	Intermediate	Low
20	0.84	0.93	0.88*
15	0.86	0.88	0.95
10	0.93	0.88	0.94

*Data only includes samples taken during operation at 0.5 mg/L DO

4.2 Industrial Bench-Scale Treatment System

The industrial bench-scale system was operated in a laboratory located at the Eastman Chemical Company WWTP (Kingsport, TN) for approximately 6 months beginning June 4, 2001. The system was fed with wastewater collected from the neutralization tank at the full-scale WWTP. Four operational phases were tested at room temperature. DO and influent ammonia loading were assessed in each of four trains operating at different SRTs (Table 4-51). Two DO levels, high (≈3.0 mg/L) and low (≈0.5 mg/L), and two influent ammonia concentration ranges, low (original wastewater from the full-scale WWTP without supplemental ammonia addition) and high (influent was augmented with ammonia), were studied during operation of the industrial wastewater activated sludge system.

Table 4-51. Operational Dates (Phases 1, 2, 3, and 4) for the Industrial Bench-scale System.

	Phase 1	Phase 2	Phase 3	Phase 4
	High DO & Low NH_4^+	High DO & High NH_4^+	Low DO& High NH_4^+	Low DO & Low NH_4^+
20-day SRT	6/4/01~8/10/01	8/11/01~9/22/01	9/23/01~11/2/01	12/5/01~12/25/01
10-day SRT	6/4/01~8/10/01	8/11/01~9/22/01	9/23/01~11/2/01	12/5/01~12/25/01
5-day SRT	6/4/01~8/10/01	8/11/01~9/22/01	9/23/01~11/2/01	12/5/01~12/25/01
2-day SRT	6/4/01~8/10/01	8/11/01~9/22/01	9/23/01~11/2/01	12/5/01~12/25/01

During Phases 1 and 2, DO levels were maintained at approximately 3 mg/L (Figure 4-39) using a DO control system. During Phases 3 and 4, DO levels were controlled at 0.5 mg/L to assess the impact of low DO concentration on the effectiveness of nitrification.

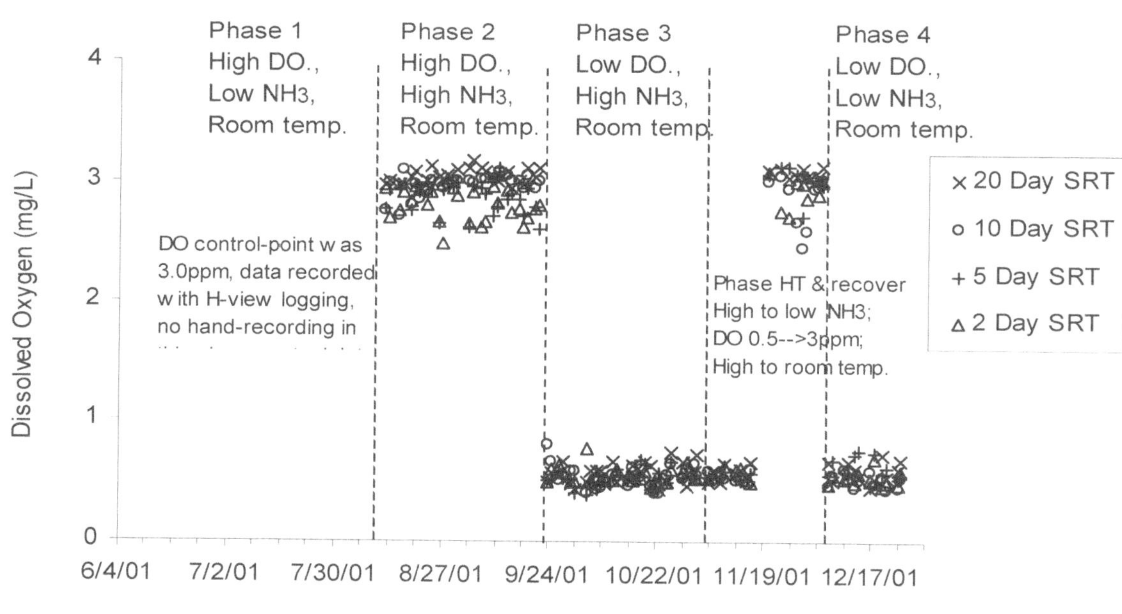

Figure 4-39. DO Profile During the Four Operating Phases of the Industrial Bench-scale System.

Phase 1 operation included treatment of industrial wastewater (unamended) collected from the WWTP. During Phase 2, ammonium chloride (NH_4Cl) was added to the wastewater to increase the influent nitrogen concentration. This was done after it became apparent that most of the influent ammonia in the unamended wastewater (Phase 1) was used to generate new bacterial cells, leaving little nitrogen for the autotrophic nitrifying bacteria. Therefore, to study nitrification, the influent ammonia concentration was approximately doubled through supplemental addition of NH_4Cl during Phases 2 and 3. During Phase 4, researchers assessed nitrification treatment performance under low ammonia (unamended wastewater) and low DO levels. Profiles of temperature and pH during the four test phases are shown in Figure 4-40.

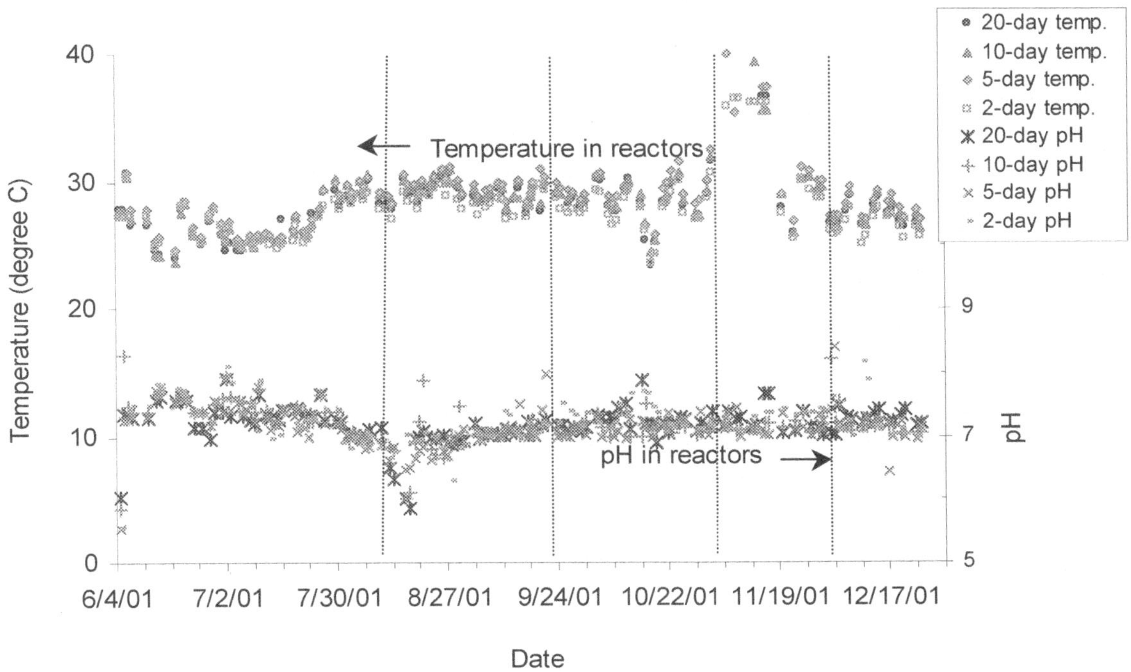

Figure 4-40. Temperature and pH Profiles During the Four Operating Phases of the Industrial Bench-scale System.

4.2.1 Mixed Liquor Volatile and Effluent Suspended Solids

The industrial bench-scale system was initially seeded with activated sludge taken from the solids recycle line of the full-scale WWTP on June 4, 2001. The SRT of each train was controlled via daily wasting of biosolids (MLSS) from each treatment train. Figure 4-41 shows MLVSS concentrations in each phase of four trains operating at different SRTs. In all phases, the MLVSS concentration was dependent on SRT; higher SRTs resulted in higher MLVSS levels. Additionally, MLVSS concentrations in the first-stage reactors of each train were higher than those in the second- and third-stage reactors. This is because return sludge was pumped back to the first-stage reactors. In addition, heterotrophic bacteria proliferated in the first and second stages because these reactors received 80% of the incoming COD. The influent flow (and therefore the COD feed) was split 50%: 30%: 20% for stages 1, 2, and 3, respectively, resulting in higher loadings to the earlier stages. Since each SRT train was defined as a system in subsequent discussions of COD and nitrification performance, only average MLVSS concentrations for each SRT train are reported (Table 4-52).

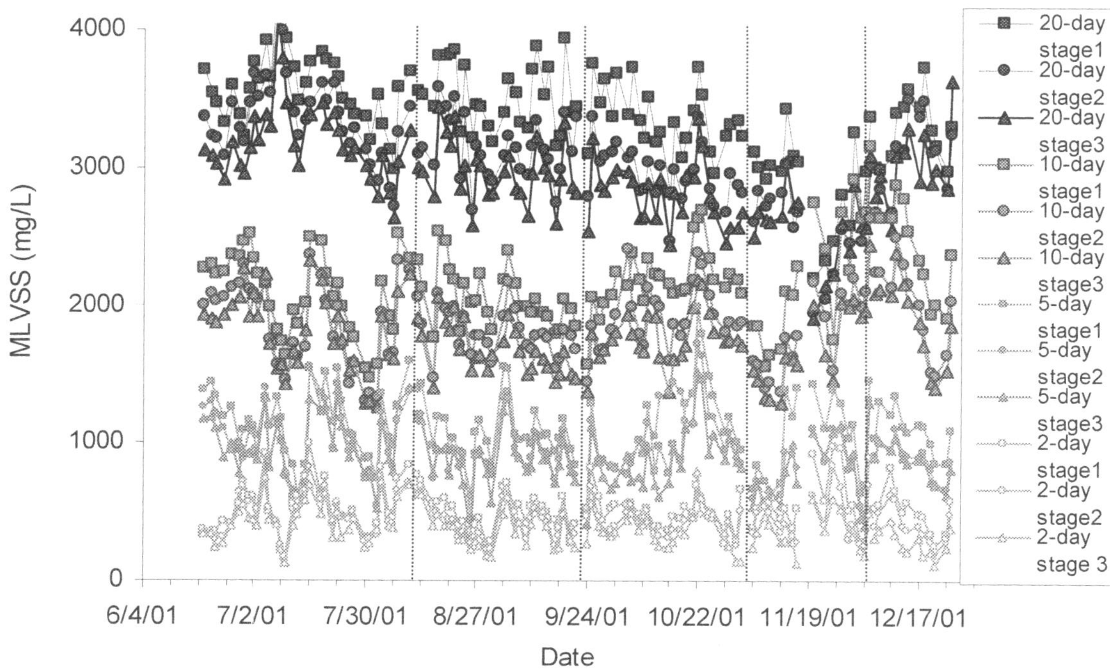

Figure 4-41. MLVSS Concentrations in Each of the 12 Reactor Units that Formed the Industrial Bench-scale System.

Table 4-52. MLVSS Concentrations (±SD) (mg/L).

	Phase 1	Phase 2	Phase 3	Phase 4
	High DO & Low NH_4^+	High DO & High NH_4^+	Low DO& High NH_4^+	Low DO & Low NH_4^+
20-day	3378 (±288)	3207 (±214)	3034 (±213)	3130 (±242)
10-day	1920 (±299)	1844 (±179)	1954 (±205)	2096 (±322)
5-day	1053 (±240)	955 (±199)	1012 (±264)	940 (±130)
2-day	485 (±171)	427 (±129)	441 (±111)	395 (±118)

The N- and P-content of the mixed liquor solids were determined for samples collected from both the bench-scale and the full-scale systems on September 4, 2001 (Table 4-53). The N-content was between 7.04% and 7.71% for the bench-scale system, somewhat higher than that in the full-scale system. The nitrogen content in biosolids was used to calculate the amount of nitrogen in the waste activated sludge (WAS) removed from each train when evaluating the overall fate of nitrogen in the industrial bench-scale system.

Table 4-53. Nitrogen and Phosphorus Content of the Biological Sludge.

	Bench-scale				Full-scale
	2-day SRT	5-day SRT	10-day SRT	20-day SRT	13.3-day SRT
P-content	1.08%	1.16%	1.27%	1.29%	0.91%
N-content	7.15%	7.71%	7.32%	7.04%	6.25%

Figure 4-42 shows ESS data over time for the industrial system. The 2-day treatment train had the highest ESS levels during the Phase 1 operation. However, ESS concentrations decreased to levels found in other SRT reactors over time. Average ESS levels during Phase 1 operation are reported in Table 4-54.

During Phase 2, the ESS levels in the 20-, 10-, and 5-day SRT trains were high due to inconsistent effluent flow. Although the effluent channels were cleaned daily, scum in the reactors would sometimes block the effluent channel, breaking through only after the water level (and pressure) increased. This break-through would occasionally cause mixing of solids in the clarifiers, resulting in high ESS concentrations.

During Phase 3, average ESS levels at each SRT were similar, although effluent flow problems continued to occur, especially in the 20-day SRT system. Following Phase 3, heating coils were used to increase the temperature in each treatment train. A lot of scum started to accumulate, and on November 17, 2001, the effluent flow from the 20-day SRT train was completely

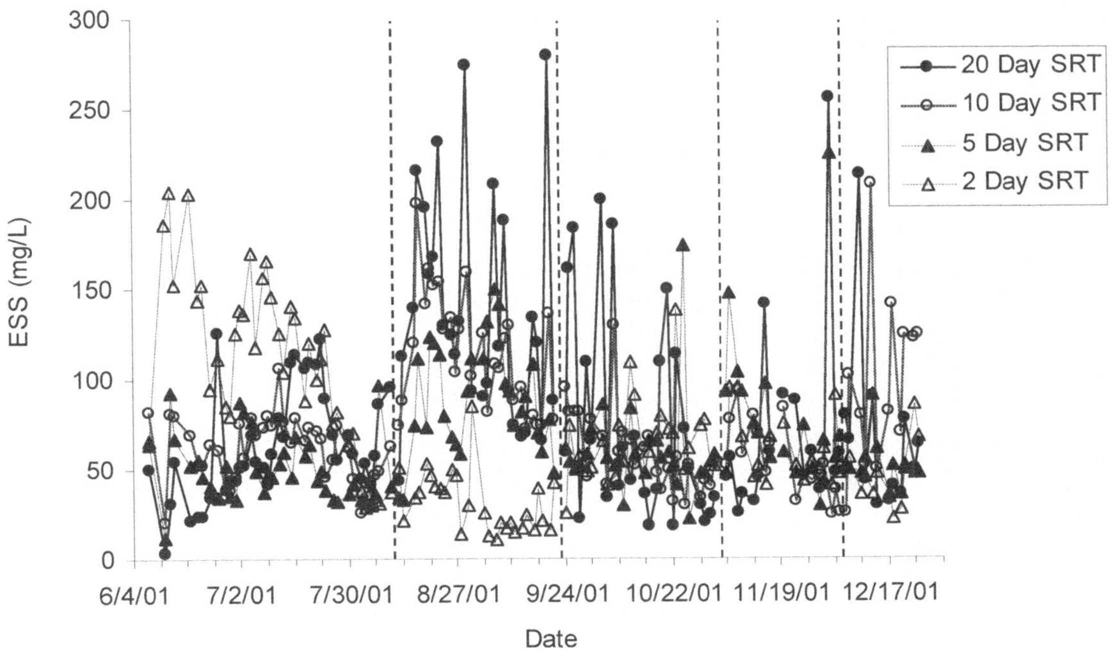

Figure 4-42. ESS Concentration in the Industrial Bench-scale System.

Table 4-54. ESS Concentration (±SD) (mg/L).

	Phase 1	Phase 2	Phase 3	Phase 4
	High DO & Low NH_4^+	High DO & High NH_4^+	Low DO& High NH_4^+	Low DO & Low NH_4^+
20-day	67 (±30)	137 (±63)	75 (±58)	69 (±50)
10-day	61 (±17)	115 (±33)	60 (±23)	99 (±49)
5-day	49 (±17)	89 (±31)	60 (±27)	52 (±15)
2-day	105 (±45)	32 (±17)	65 (±23)	47 (±18)

blocked causing the mixed liquor to overflow from the top of the reactors. This caused unusually high loss of sludge. The MLVSS concentration changed from 3040 (on November 16) to 2200 (on November 20) in Phase 1, from 2680 mg/L to 1940 mg/L in Phase 2 and from 2750 mg/L to 1900 mg/L in Phase 3. Because of this unusual sludge loss, the test utilizing high temperature was stopped. The heating coils were removed on November 20, 2001, the DO levels were changed to 3 mg/L, and sludge wasting was stopped for 2 weeks until (December 2, 2001) in order to let the sludge concentration recover. The SRT value in this train was not evaluated during this period since the amount of sludge loss was unknown.

ESS levels were highly variable in the 10-day SRT reactor during Phase 4. ESS levels in the other three reactors typically ranged between 25 mg/L and 75 mg/L during Phase 4.

4.2.2 Chemical Oxygen Demand Treatment Performance

Figure 4-43(A) shows COD concentrations of influent and effluent samples collected from the four trains. Influent wastewater COD concentrations were highly variable during the entire period, although average levels were highest in Phase 1 and lowest in Phase 4 (Table 4-55). However, COD removal performance was fairly stable during all test phases (Figure 4-43(B). For most of the Phase 1, 2, and 4, effluent COD concentrations were lower than 100 mg/L, and removal efficiencies were higher than 90% in all four trains. During Phase 3 (DO=0.5 mg/L), average removal efficiencies were between 85% and 90% for each SRT, significantly lower (α=0.05) than those measured during Phase 2 (DO=3.0 mg/L). Thus, the lower DO level did impact COD removal performance in the industrial bench-scale system.

Table 4-55. Average (±SD) Influent and Effluent COD Concentrations (mg/L) with Removal Efficiencies (%).

	Phase 1	Phase 2	Phase 3	Phase 4
	High DO & Low NH_4^+	High DO & High NH_4^+	Low DO& High NH_4^+	Low DO & Low NH_4^+
Influent	1084 (±157)	818 (±194)	844 (±293)	468 (±92)
20-day Effluent	43 (±21) (96.0%*)	25 (±13) (96.8%)	73 (±26) (89.9%)	16 (±13) (96.3%)
10-day Effluent	49 (±23) (95.3%)	34 (±25) (95.6%)	103 (±35) (86.1%)	29 (±28) (93.0%)
5-day Effluent	63 (±56) (94.0%)	40 (±24) (94.9%)	98 (±31) (86.8%)	26 (±18) (94.3%)
2-day Effluent	87 (±47) (91.9%)	39 (±15) (94.9%)	98 (±33) (86.4%)	38 (±36) (92.0%)

Average COD removal performance in the 20-day train was always slightly better than that in lower SRT trains. Although treatment level differences between the 20-, 10- and 5-day SRT reactors during operation at each phase were small, the average COD removal efficiency of the 20-day train was significantly different (higher) (α=0.05) than that of the 2-day train.

4.2.3 Nitrification Performance

TKN concentrations of influent and effluent samples were measured twice per week over the treatability study (Table 4-56). Additionally, influent and effluent ammonia, nitrite, and nitrate were measured each day. Removal efficiencies were, on average, >90% for the 20-, 10- and 5-day SRT reactors based on TKN data. Nitrogen removal efficiencies in the 2-day SRT system ranged from 50% to 90%.

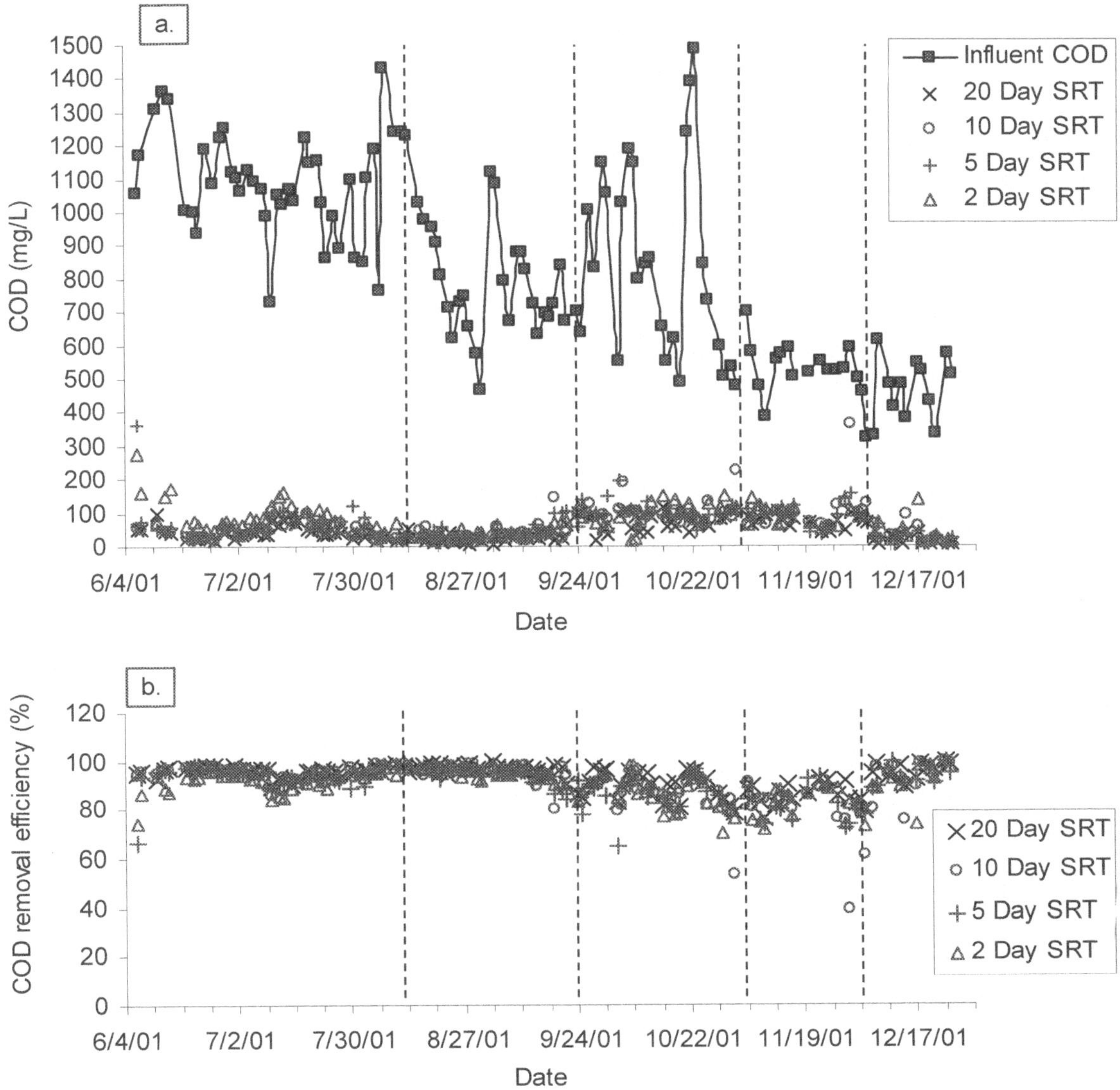

Figure 4-43. Influent and Effluent COD Concentrations (A) and COD Removal Efficiency (B) in the Industrial Bench-scale System.

Table 4-56. Average (±SD) Influent and Effluent TKN Concentrations (mg/L as N) with Removal Efficiencies (%).

	Phase 1	Phase 2	Phase 3	Phase 4
	High DO & Low NH_4^+	High DO & High NH_4^+	Low DO& High NH_4^+	Low DO & Low NH_4^+
Influent	21.4 (±4.8)	54.2 (±9.1)	53.3 (±11.0)	26.5 (±12.1)
20-day Effluent	1.3 (±0.8) (94.1%)	0.7 (±0.4) (98.6%)	1.1 (±1.1) (98.3%)	1.1 (±0.5) (95.2%)
10-day Effluent	1.6 (±0.6) (92.1%)	2.4 (±1.8) (95.0%)	0.5 (±0.5) (99.0%)	1.4 (±1.0) (94.1%)
5-day Effluent	2.2 (±1.3) (89.9%)	1.5 (±1.2) (97.4%)	2.3 (±2.3) (95.9%)	0.8 (±0.5) (96.6%)
2-day Effluent	3.0 (±1.1) (85.6%)	5.2 (±5.2) (91.0%)	27.8 (±20.0) (50.8%)	2.6 (±2.0) (88.8%)

Influent and effluent ammonia concentrations over time are shown in Figure 4-44A. Additionally, ammonia and nitrite oxidation efficiencies are showed in Figures 4-44B and 4-44C, respectively. Ammonia oxidation efficiency was evaluated using the following equation:

$$AOE(\%) = \frac{\{Eff_{NO_2^-} + Eff_{NO_3^-} - Inf_{NO_2^-} - Inf_{NO_3^-}\}}{\{Eff_{NH_4^+} + Eff_{NO_2^-} + Eff_{NO_3^-} - Inf_{NO_2^-} - Inf_{NO_3^-}\}} \times 100\% \qquad (4.2.3\text{-}1)$$

where: AOE(%) = ammonia oxidation efficiency
$Eff_{NO_2^-}$ = effluent nitrite (mg/L as N)
$Eff_{NO_3^-}$ = effluent nitrate (mg/L as N)
$Inf_{NO_2^-}$ = influent nitrite (mg/L as N)
$Inf_{NO_3^-}$ = influent nitrate (mg/L as N)
$Eff_{NH_4^+}$ = effluent ammonia (mg/L as N)

When $(Eff_{NO_2^-} + Eff_{NO_3^-}) \leq (Inf_{NO_2^-} + Inf_{NO_3^-})$ or $(Eff_{NO_2^-} + Eff_{NO_3^-}) - (Inf_{NO_2^-} + Inf_{NO_3^-}) \leq 0$, it was assumed that the ammonia oxidation efficiency was zero (i.e., ammonia oxidation did not occur). The nitrite oxidation efficiency was computed in a similar manner.

$$NOE(\%) = \frac{\{Eff_{NO_3^-} - Inf_{NO_3^-}\}}{\{Eff_{NH_4^+} + Eff_{NO_2^-} + Eff_{NO_3^-} - Inf_{NO_3^-}\}} \times 100\% \qquad (4.2.3\text{-}2)$$

where: NOE(%) = nitrite oxidation efficiency

When $Eff_{NO_3^-} \leq Inf_{NO_3^-}$ or $Eff_{NO_3^-} - Inf_{NO_3^-} \leq 0$, it was assumed that nitrite oxidation efficiency was zero (i.e. nitrite oxidation did not occur).

For analysis and discussion purposes, steady state for ammonia and nitrite treatment performance during each operating phase was determined as the period in which the oxidation efficiencies were fairly stable (i.e., obvious shock loading or transitional periods were not included). Average ammonia- and nitrite-oxidation efficiencies during steady state are summarized in Table 4-57.

Table 4-57. %Ammonia-Oxidation Efficiency (AOE) and %Nitrite-Oxidation Efficiency (NOE) During Steady State.

	Phase 1		Phase 2		Phase 3		Phase 4	
	High DO & Low NH_4^+		High DO & High NH_4^+		Low DO& High NH_4^+		Low DO & Low NH_4^+	
	AOE	NOE	AOE	NOE	AOE	NOE	AOE	NOE
20-day	86.1	96.2	99.7	99.7	95.8	96.3	99.8	99.8
10-day	60.4	78.4	99.6	99.6	98.4	98.5	99.3	99.3
5-day	<0.1	7.6	99.5	99.6	94.8	93.9	96.8	97.0
2-day	<0.1	0.5	90.6	75.6	2.3	2.0	88.1	36.7

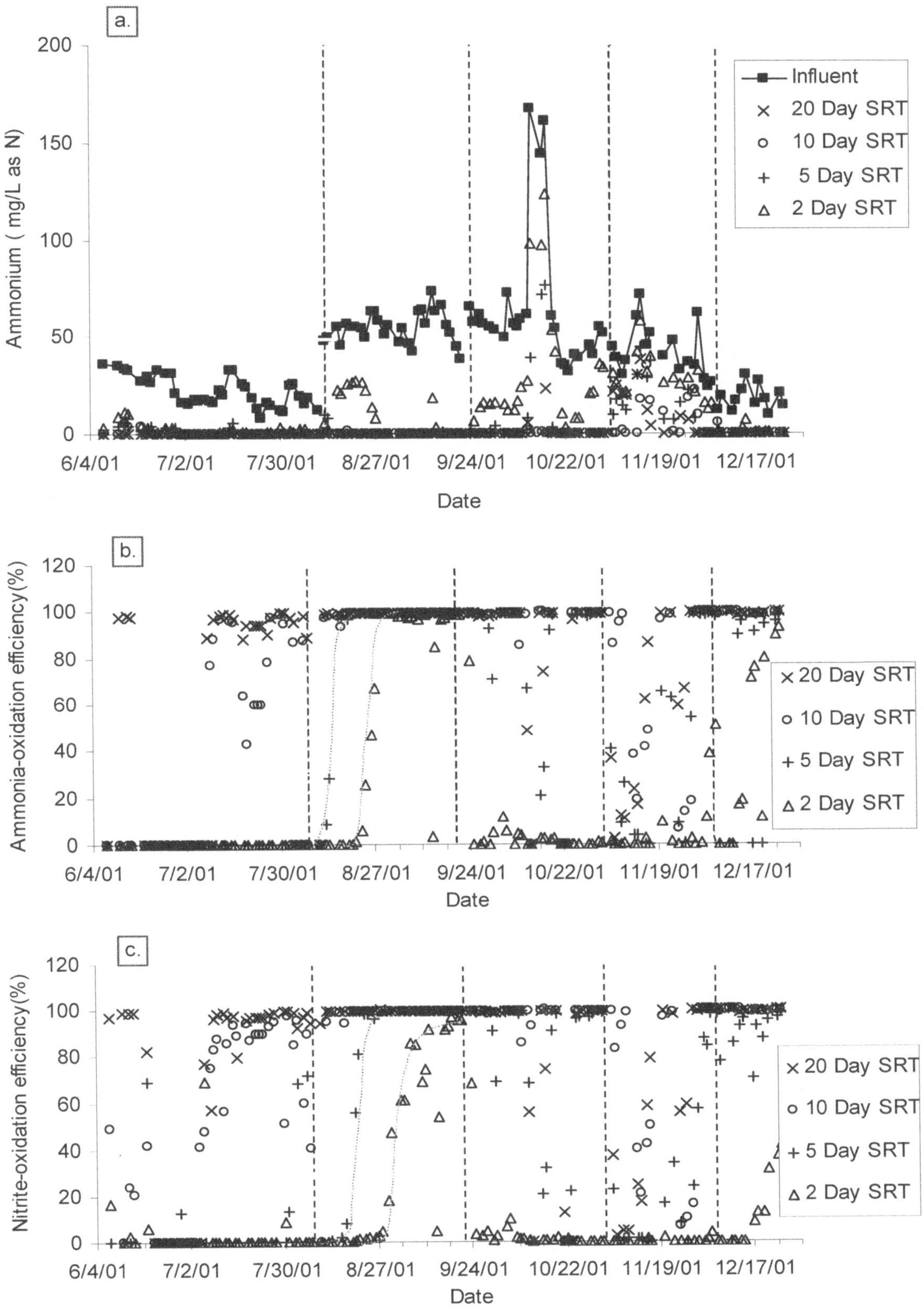

Figure 4-44. Influent and Effluent Ammonia Concentrations (A), Ammonia-Oxidation Efficiency (B), and Nitrite-Oxidation Efficiency (C) in the Industrial Bench-scale System.

During Phase 1 operation, the average influent ammonia concentration was 19 mg/L (as N) while average effluent concentrations were below 1.0 mg/L in all four trains. However, most of the influent nitrogen was likely utilized during growth and synthesis activities by heterotrophic bacteria. Nitrogen required for heterotrophic growth was estimated using a COD: nitrogen ratio of 30:1 (developed from operation of the full-scale industrial WWTP). Little nitrogen was available for nitrifying bacteria during Phase 1 (Table 4-58).

Table 4-58. Estimated Ammonia-N Available for Autotrophic Ammonia-Oxidizing Bacteria.

	Phase 1	Phase 2	Phase 3	Phase 4
Average COD removed by heterotrophic bacteria (mg/L)	1029	783	749	442
Nitrogen required for heterotrophic bacterial growth (mg/L as N)	34.3	26.1	25.0	14.7
Average influent nitrogen (mg/L as N)	18.9	54.6	52.8	19.2
Nitrogen available for autotrophic AOB (mg/L as N)	0	28.5	27.8	4.5

Average influent and effluent concentrations of all nitrogen species evaluated during steady-state operation of each system during each test phase are summarized in Table 4-59. Figure 4-45 shows the ammonia conversion profile during Phase 1 (low ammonia, high DO) operation. This figure shows that most of the influent ammonia-N was converted to biomass-N (based on new growth equal to wasted biomass during steady-state operation). The 2- and 5-day SRT effluents contained nitrite plus nitrate levels that were similar to influent ammonia concentrations (on a nitrogen basis) most of time. Since ammonia- and nitrite-oxidation efficiencies for the 2- and 5-day trains were very low during Phase 1 (Table 4-57), it was concluded that the nitrification was not a significant reaction at these SRTs because of the limited nitrogen available. For the 10- and 20-day SRT trains, some nitrite and nitrate were detected in the effluent (Figures 4-45D and 4-45E) indicating that microbial growth requirements were satisfied and excess nitrogen was available to the nitrifying population. Ammonia- and nitrite-oxidation efficiencies for the longer SRTs were much higher than those found in the shorter SRT systems (Table 4-57).

Figure 4-46 shows ammonia conversion during Phase 2 (high ammonia, high DO) operation. Steady-state concentrations are summarized in Table 4-59. The average influent ammonia concentration was increased from 19 mg/L to 55 mg/L (as N) via addition of ammonium chloride. Although a large portion of the influent nitrogen was taken up by the biomass, significant nitrification occurred at all SRTs. Bacteria in the 20- and 10-day systems quickly converted the excess ammonia to nitrate (no lag phase). The 5-day SRT system required approximately 1 week to achieve nitrification, while the 2-day SRT system required 2.5 weeks. A transition period was noted in the 2-day system (Figure 4-46B). For approximately the first 10 days, effluent ammonia levels were equal to the supplemental levels added to the influent. Nitrate concentrations in the effluent were at or below the detection limit. However, nitrite levels began to rise after 1 week and a corresponding drop in effluent ammonia level was measured. Ammonia-oxidation efficiency started to increase gradually during this period. Nitrite build-up continued for about 5 days before levels began to drop. Ultimately, effluent ammonia and nitrite reached low levels and nitrate was measured in

Table 4-59. Average Concentrations of Target Nitrogen Forms in Influent and Effluent.

		Parameter	2-day SRT	5-day SRT	10-day SRT	20-day SRT
Phase 1	Influent	TKN			21.4 (±4.8)	
		NH_4^+			18.9 (±6.9)	
		NO2-			2.8 (±0.5)	
		NO3-			0.2 (±0.2)	
		Total Inf. N			24.4	
	Effluent	TKN	3.0 (±1.1)	2.2 (±1.3)	1.6 (±0.6)	1.3 (±0.8)
		NH_4^+	0.8 (±0.8)	0.8 (±0.8)	<0.1	<0.1
		NO2-	0.2 (±0.2)	0.4 (±0.4)	0.6 (±0.6)	<0.1
		NO3-	<0.1	0.2 (±0.2)	3.5 (±1.6)	7.8 (±5.0)
		WAS N	23.3 (±9.4)	21.5 (±6.1)	19.2 (3.4)	17.0 (1.3)
		Total Eff. N	26.5	24.3	24.9	26.1
	N recovery	(%)	108.6	99.6	102.0	107.0
Phase 2	Influent	TKN			54.2 (±9.1)	
		NH_4^+			54.6 (±10.2)	
		NO2-			1.5 (±0.3)	
		NO3-			0.1 (±0.1)	
		Total Inf. N			55.8	
	Effluent	TKN	5.2 (±5.2)	1.5 (±1.2)	2.4 (±1.8)	0.7 (±0.4)
		NH_4^+	1.7 (±1.7)	<0.1	<0.1	<0.1
		NO2-	2.3 (±2.2)	<0.1	<0.1	<0.1
		NO3-	11.0 (±4.5)	26.2 (±5.3)	31.6 (±6.8)	34.2 (±6.1)
		WAS N	21.0 (±5.8)	20.0 (±4.1)	18.5 (±1.6)	16.8 (±1.1)
		Total Eff. N	39.5	47.7	52.5	51.7
	N recovery	(%)	70.8	85.5	94.1	92.7
Phase 3	Influent	TKN			53.3 (±11.0)	
		NH_4^+			52.8 (±4.2)	
		NO2-			0.8 (±0.7)	
		NO3-			0.1 (±0.1)	
		Total Inf. N			54.2	
	Effluent	TKN	20.7 (±9.5)	2.3 (±2.3)	0.5 (±0.5)	1.1 (±1.1)
		NH_4^+	18.9 (±8.4)	1.0 (±1.0)	0.2 (±0.2)	0.5 (±0.5)
		NO2-	0.5 (±0.4)	0.2 (±0.1)	<0.1	<0.1
		NO3-	0.4 (±0.4)	18.7 (±6.4)	20.5 (±7.7)	16.4 (±8.1)
		WAS N	18.6 (±4.2)	19.1 (±3.2)	20.4 (±1.3)	15.3 (±0.8)
		Total Eff. N	40.2	40.3	41.4	32.8
	N recovery	(%)	74.2	74.4	76.4	60.5
Phase 4	Influent	TKN			21.5 (±5.7)	
		NH_4^+			19.2 (±6.6)	
		NO2-			0.4 (±0.2)	
		NO3-			<0.1	
		Total Inf. N			21.9	
	Effluent	TKN	2.6 (±2.0)	0.8 (±0.5)	1.4 (±1.0)	1.1 (±0.5)
		NH_4^+	1.5 (±1.5)	<0.1	<0.1	<0.1
		NO2-	1.5 (±0.9)	<0.1	<0.1	<0.1
		NO3-	0.2 (±0.2)	1.4 (±0.9)	8.6 (±3.8)	11.2 (±4.3)
		WAS N	16.4 (±5.4)	18.8 (±2.9)	17.6 (±1.6)	16.3 (±1.3)
		Total Eff. N	20.7	21.0	27.6	28.6
	N recovery	(%)	94.5	95.9	126.0	130.6

Total Inf. N = Influent (TKN+ NO_2^-+ NO_3^-); **Total Eff. N =** Effluent (TKN+ NO_2^-+ NO_3^-+Waste N); **N recovery (%) =** (Total Inf. N) ×100% / (Total Eff. N)

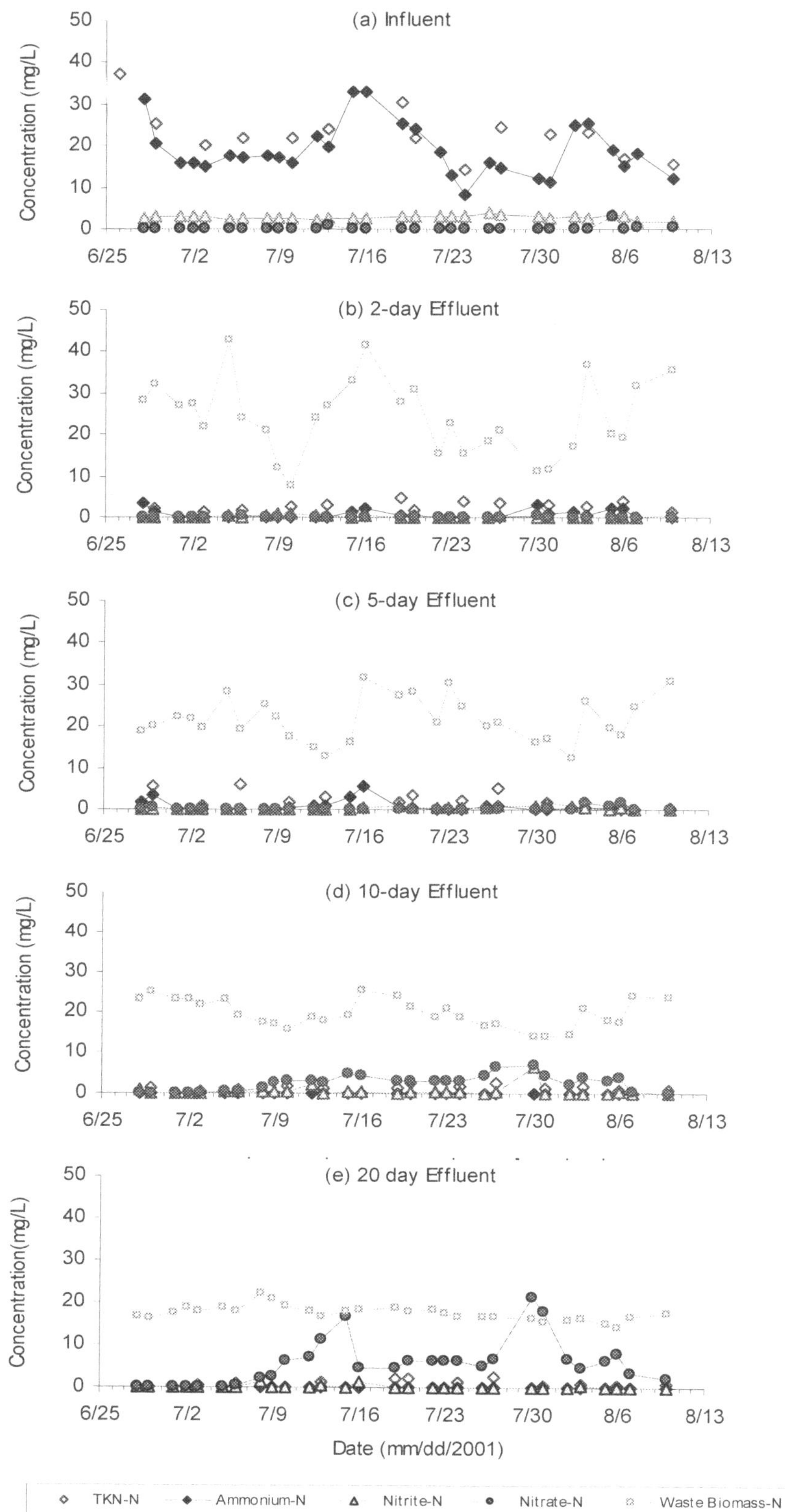

Figure 4-45. Concentrations of Target Nitrogen Species in Influent (A) and Effluent (B, C, D, E) During Phase 1.

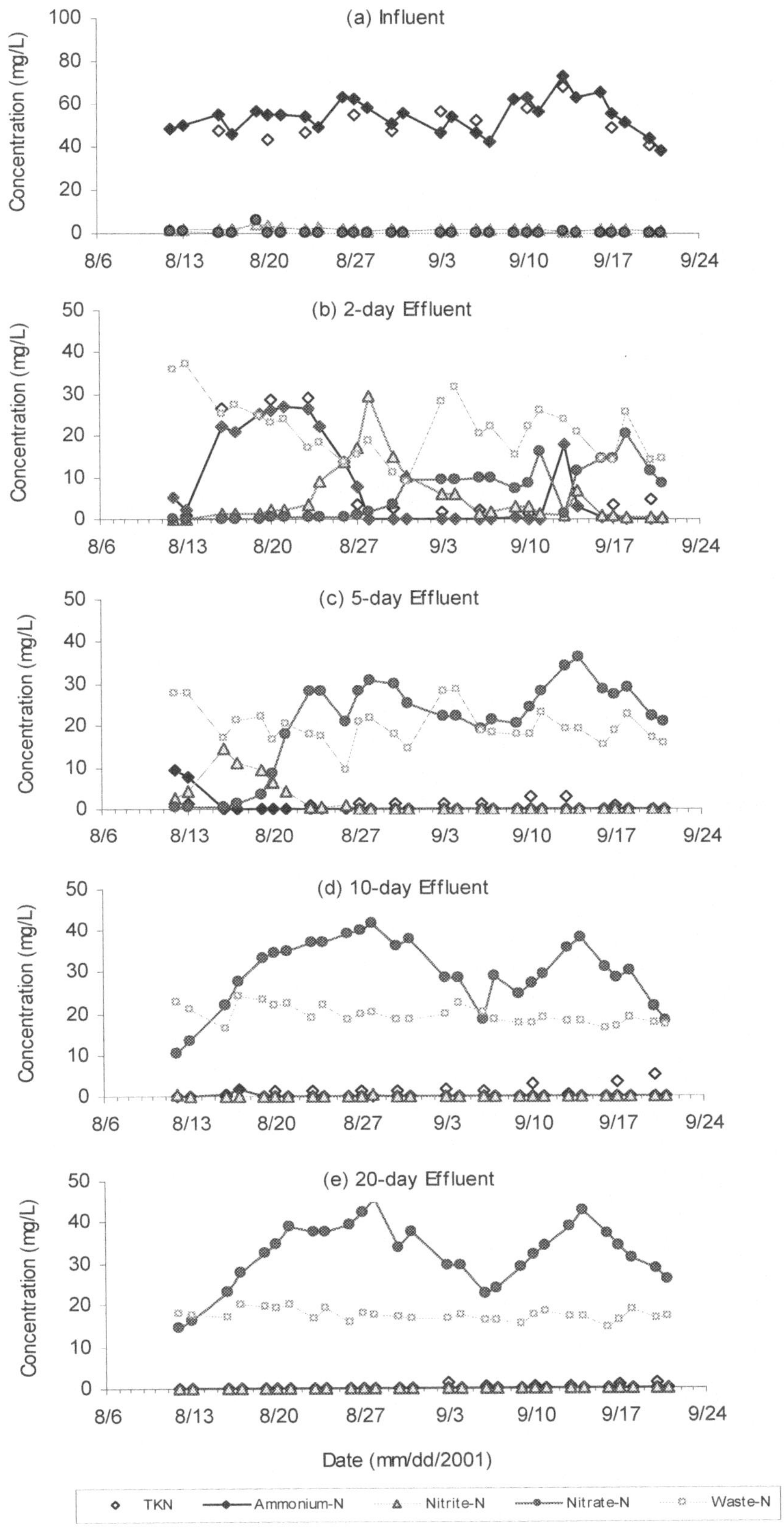

Figure 4-46. Concentrations of Target Nitrogen Species in Influent (A) and Effluent (B, C, D, E) During Phase 2.

the effluent. It is believed that the population of ammonia- and nitrite-oxidizers, or their activity, changed as a result of the increased ammonia loading.

Figure 4-47 shows ammonia conversion throughout Phase 3. During this phase, DO was lowered to 0.5 mg/L while ammonia supplementation continued. The low DO level impacted ammonia- and nitrite-oxidation performance, especially in the 2-day SRT system where ammonia- and nitrite-oxidation efficiencies significantly decreased. The 5-, 10- and 20-day SRT trains continued to achieve good nitrification. However, effluent nitrate levels were lower than those measured during Phase 2. Less than 80% of the nitrogen in the influent was recovered from the effluent. This suggests that other end products that were not measured in this study may have been produced under low DO conditions. Gaseous products from denitrification, such as N_2 or N_2O, may account for the missing nitrogen mass.

It should be noted that a shock-loading period (discharge of NH_4NO_3 into the full-scale WWTP) occurred during Phase 3. High levels of ammonia and nitrate were detected in the influent for the industrial bench-scale system on October 11, 2001, and continued for 6 days. Initially, the 20- and 10-day SRT systems effectively treated the shock load. Most of the effluent-N was in the nitrate form with little ammonia present in the effluent of these SRT reactors. Production of nitrate remained somewhat constant in the 5-day SRT system during the shock loading. The additional ammonia passed through the system untreated.

The ammonia- and nitrite-oxidation efficiencies during high DO (Phase 2), low DO (Phase 3), and ammonia shock loading periods (in Phase 3) were compared (Figure 4-48A and B). The ammonia- and nitrite-oxidation efficiencies of the 2-day train were the lowest of the four SRTs at high DO (Phase 2) and significantly decreased as the DO level was dropped (Phase 3). Ammonia- and nitrite-oxidation efficiencies in the 5-, 10-, and 20-day trains remained high when the DO was dropped to 0.5 mg/L. However, a significant decrease occurred during the shock ammonia loading. The impact of low DO level and ammonia shock loading on ammonia- and nitrite-oxidation efficiencies was evaluated using a t-test (α= 0.05). Only the 2-day SRT system showed that the ammonia-oxidation efficiency during low DO phase (Phase 3) was significantly different (lower) than that during high DO phase (Phase 2). Although the low DO level did not significantly affect nitrification performance in the 20-, 10-, and 5-day SRT systems, other impacts of low DO may have occurred. Nitrogen conversion mechanisms, other than the ammonia to nitrate pathway, can occur under low DO conditions (i.e., denitrification). During this study, average influent ammonia (and COD) levels were nearly equal at both low and high DO phases; however, effluent nitrate concentrations during Phase 3 (low DO) were lower than those measured during Phase 2 (high DO). Statistical testing to assess differences in ammonia- and nitrite-oxidation performance under low and high DO levels did not consider possible denitrification reactions. Therefore, although low DO operation (Phase 3) did not effect ammonia removal performance in 20-, 10-, and 5-day SRT trains, production of nitrate was impacted.

Interestingly, nitrate in the 20- 10- and 5-day SRT systems decreased to below detection levels after the shock loading and remained there for 3 days. The concentration then slowly rose over time to levels measured before introduction of the shock load. In the 2-day SRT train during Phase 3 operation, little, if any, nitrification occurred prior to, during or after shock loading.

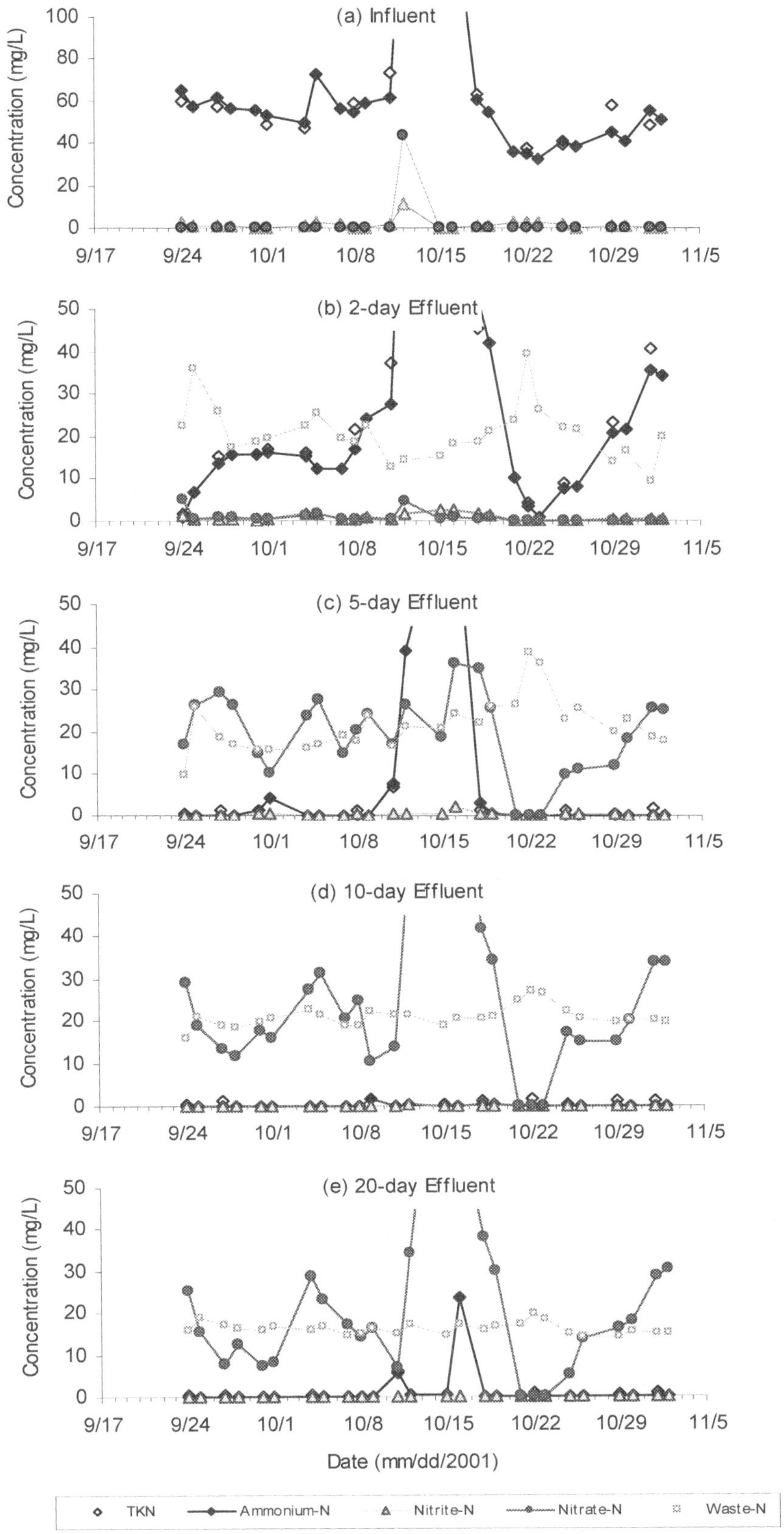

Figure 4-47. Concentrations of Target Nitrogen Species in Influent (A) and Effluent (B, C, D, E) During Phase 3.

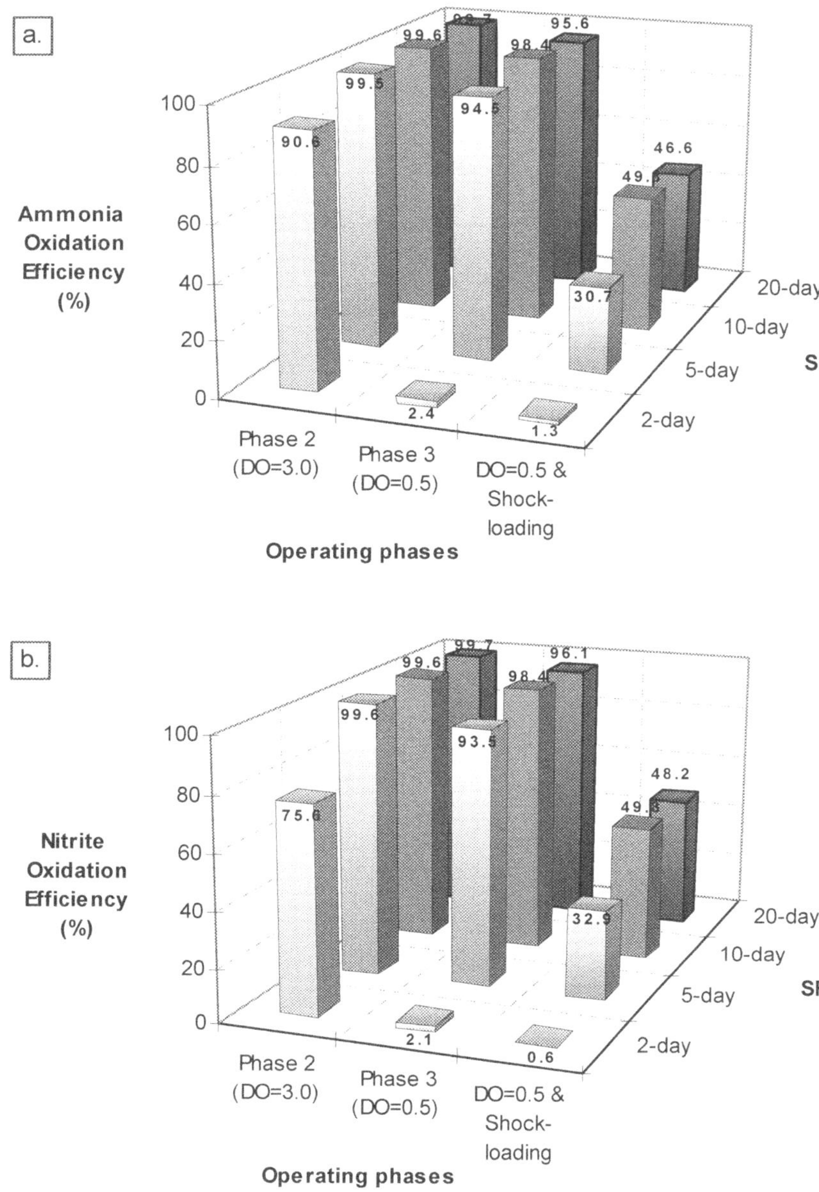

Figure 4-48. Comparison of Average Ammonia-oxidation Efficiency (A) and Nitrite-oxidation Efficiency (B) Under Different Operating Conditions (changing DO and ammonia shock loading).

Figure 4-49 shows ammonia conversion during Phase 4. In this phase, supplemental nitrogen loading was stopped, resulting in an average influent ammonia concentration of approximately 19 mg/L (as N), similar to the low ammonia level in Phase 1. However, the DO level during Phase 4 was reduced to 0.5 mg/L. Results from Phases 1 and 4 are similar in that little of the influent ammonia was released in the effluent. Biological conversion (nitrification) or cellular uptake led to the low effluent ammonia levels measured. For the 2- and 5-day SRTs, cellular uptake accounted for most of the removed ammonia, leaving little nitrogen available for nitrification. However, during Phase 4, the average influent COD level was approximately one-half of that measured during Phase 1

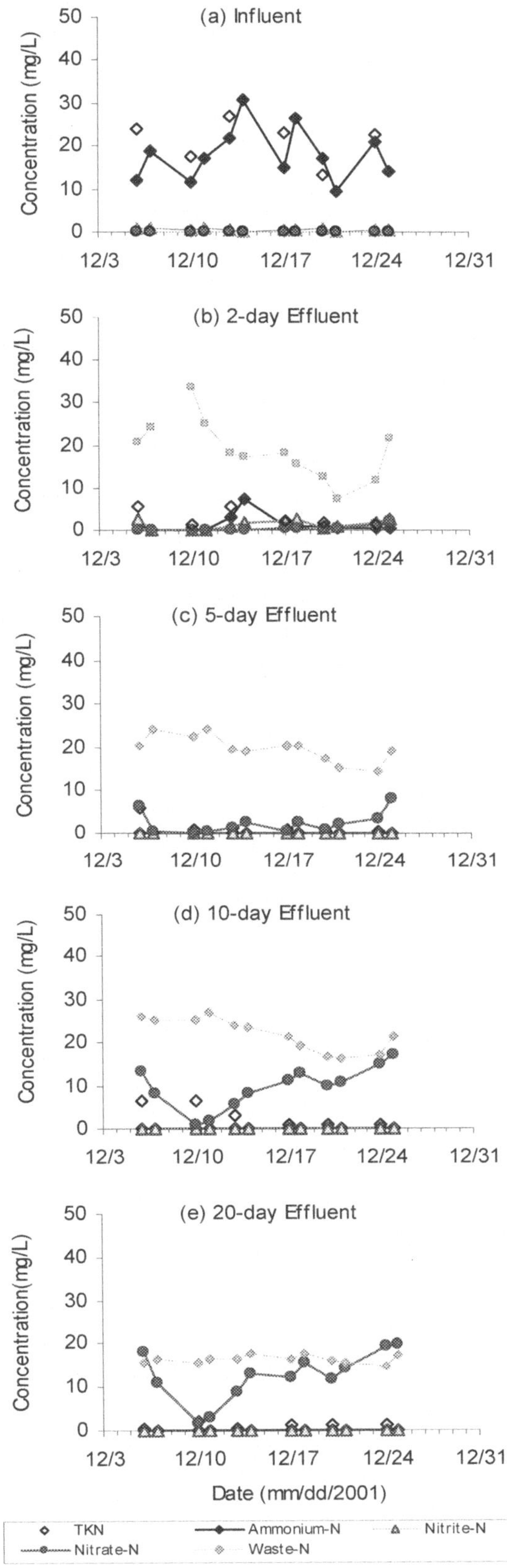

Figure 4-49. Concentrations of Target Nitrogen Species in Influent (A) and Effluent (B, C, D, E) During Phase 4.

operation resulting in less heterotrophic growth and less nitrogen uptake. Therefore, additional nitrogen was available for nitrification during Phase 4 (Table 4-58). Since little nitrate was measured in the influent, the presence of nitrate in the effluent indicated nitrification had occurred under the low ammonia-low DO conditions. However, nitrate production remained very low in the 2- and 5-day SRT systems, as cellular uptake dominated nitrogen removal. Nitrate production in the 10- and 20-day SRT reactors during Phase 4 accounted for a substantial amount of the nitrogen released from these reactors.

4.2.4 Total Bacteria (Using the 16S-rDNA Gene as the Target)

Biosolids samples taken from the industrial bench-scale system during each operating phase were also used to quantify microbial populations. Bacterial DNA was extracted from selected samples and used to assess target bacterial numbers. Three real-time PCR assays were designed and optimized for total bacteria (16S rDNA of eubacteria), AOB (*amo*A gene of *Nitrosomonas nitrosa*-type), and NOB (16S rDNA of *Nitrospira*-type).

A total of 64 industrial bench-scale biosolids samples (4 samples from each SRT train during each operating phase) were chosen for genetic analysis. The selection of samples for molecular analysis was grouped based on SRT and operating phases (different DO level and ammonia concentrations). Table 4-60 lists the sample dates chosen for genetic analysis. The molecular results were normalized to cells per liter (mixed liquor) using the following values: 3.6 copies of 16S rDNA per cell for total eubacteria, one copy of 16S rDNA per cell for *Nitrospira* and two copies of the *amoA* gene per cell for *N. nitrosa*-type AOB.

Table 4-60. Sample Dates for Molecular Analysis of Industrial Bench-scale Samples.

	Phase 1	Phase 2	Phase 3	Phase 4
	High DO & Low NH_4^+	High DO & High NH_4^+	Low DO& High NH_4^+	Low DO & Low NH_4^+
Sample Dates for Each SRT	07/16/02	08/27/02	10/04/02	12/10/02
	07/23/02	09/03/02	10/08/02	12/17/02
	07/27/02	09/10/02	10/25/02	12/20/02
	08/03/02	09/17/02	10/29/02	12/24/02

Table 4-61 presents the average total bacterial number under each operating phase during industrial bench-scale treatment. Numbers of total bacteria (cells/L) in each SRT train were somewhat constant during the four operating periods; however, the longer SRT trains generally had higher total bacteria numbers (per liter of mixed liquor) than the shorter SRT trains (Figure 4-50). Total bacteria concentrations (cells/L) increased when MLVSS levels increased in the longer SRT trains. The linear relationship depicted in Figure 4-51 indicates that higher total bacteria levels (cells/L) were found in samples with higher concentrations of MLVSS. When the total bacteria number was normalized to mass (mg) of MLVSS, the values were essentially constant at approximately 1.31×10^9 (cells/mg MLVSS) in all four SRT trains during the operating period (Figure 4-52). This suggests that the SRTs, DO levels, influent ammonia concentrations, and COD concentrations did not significantly change the normalized total eubacteria number. Because of the relatively high influent COD levels, heterotrophic bacteria should predominate the microbial population in the industrial treatment system. Thus, the number of heterotrophic bacteria should be closely aligned with the number of the total bacteria.

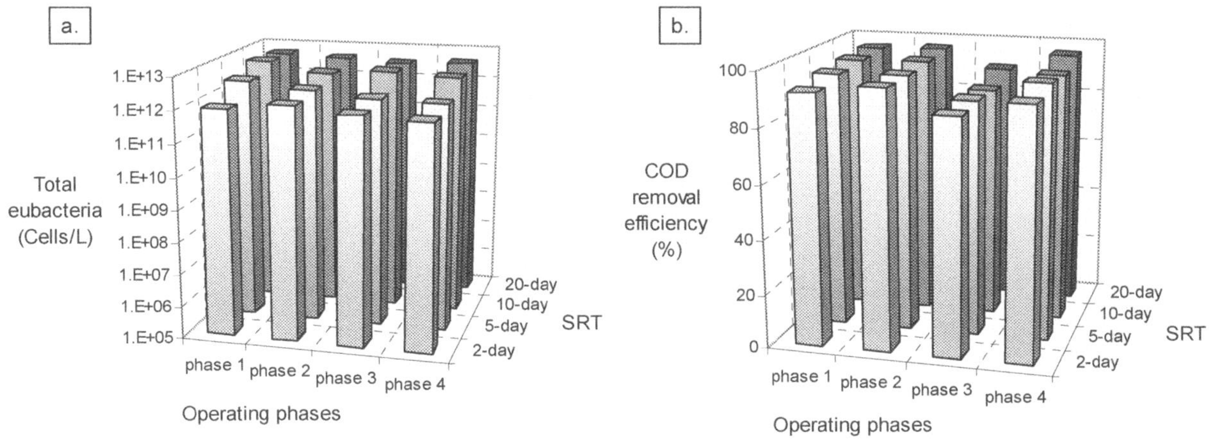

Figure 4-50. Total Eubacteria Concentration (A) Compared with the COD Removal Efficiency (B) in the Industrial Bench-scale System.

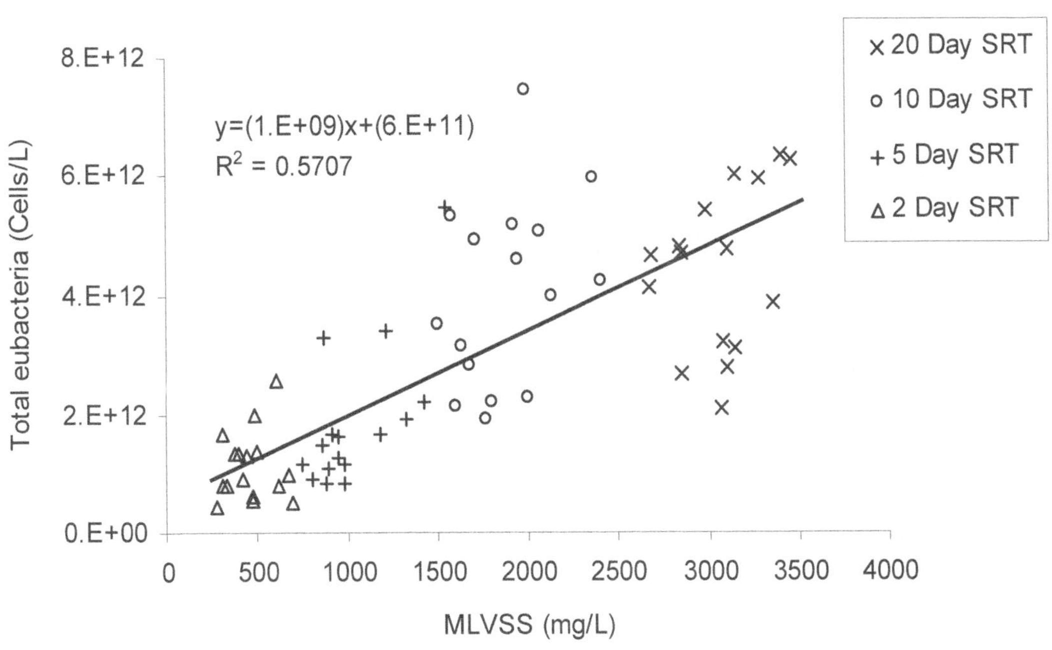

Figure 4-51. Correlation Between the MLVSS Concentration (mg/L) and the Total Eubacteria Concentration (cells/L).

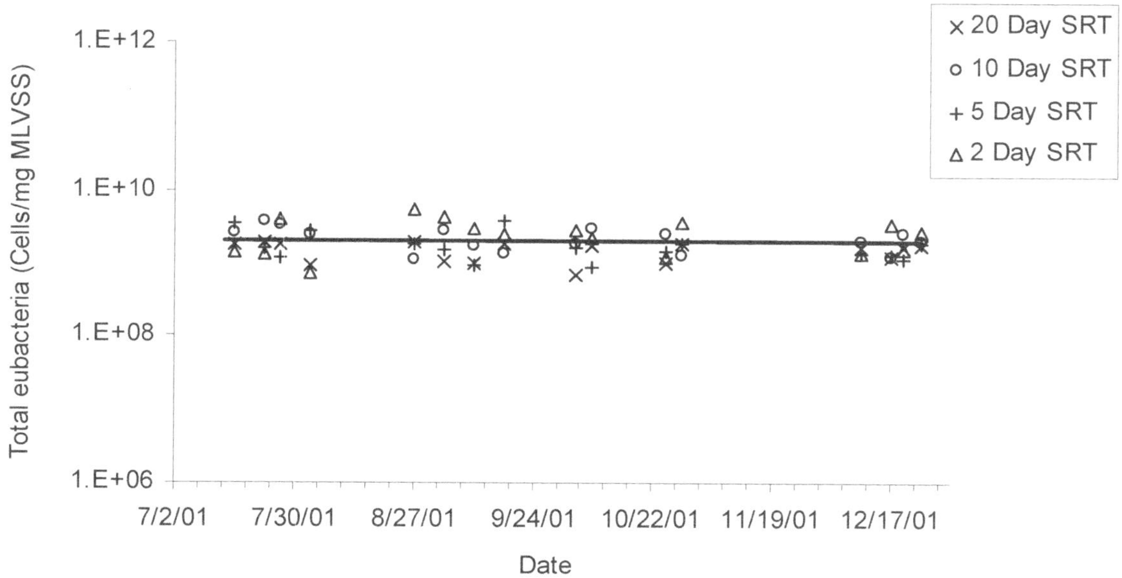

Figure 4-52. Total Eubacteria Number Normalized to Mass (mg) of MLVSS During Activated Sludge Treatment of Industrial Wastewater.

Table 4-61. Quantification of Total Bacteria (Cells/L) in the Industrial Bench-scale System Using Real-time PCR Assay to Target 16S rDNA of all Eubacteria.

	Phase 1	Phase 2	Phase 3	Phase 4
20-day	$5.36 (\pm1.73) \times 10^{12}$	$4.45 (\pm1.50) \times 10^{12}$	$3.53 (\pm1.35) \times 10^{12}$	$4.39 (\pm0.48) \times 10^{12}$
10-day	$5.84 (\pm1.23) \times 10^{12}$	$3.01 (\pm1.51) \times 10^{12}$	$4.12 (\pm1.32) \times 10^{12}$	$3.24 (\pm0.72) \times 10^{12}$
5-day	$3.06 (\pm1.85) \times 10^{12}$	$1.92 (\pm1.02) \times 10^{12}$	$1.32 (\pm0.40) \times 10^{12}$	$1.18 (\pm0.26) \times 10^{12}$
2-day	$1.01 (\pm0.68) \times 10^{12}$	$1.58 (\pm0.76) \times 10^{12}$	$1.04 (\pm0.40) \times 10^{12}$	$0.84 (\pm0.37) \times 10^{12}$

Figure 4-53 shows that COD removal efficiency was high at most total bacterial concentrations (i.e., COD removal efficiency did not significantly correlate to total bacteria concentration in the system). Even though the total bacteria concentrations (cells/L) in the 2- and 5-day trains were lower than those in the 10- and 20-day trains, COD treatment performance in the shorter SRT systems was as good as that measured in the longer SRT systems (on average). However, COD treatment performance was not as consistent in the 2- and 5-day systems as that in the 10- and 20-day systems. Longer SRTs (which yield higher total bacterial levels) seem to provide more stable operation.

4.2.5 *Nitrosomonas nitrosa*-type Ammonia-Oxidizing Bacteria (AOB)

The *Nitrosomonas nitrosa*-type bacteria were expected to be a dominant species of AOB in the industrial bench-scale system since this group was found to be the predominate AOB in the full-scale industrial WWTP. The real time-PCR assay used to quantify the *N. nitrosa*- type AOB in the full-scale system was similarly applied to quantify this organism in the bench-scale system. Results are presented in Table 4-62. In all four SRT trains, the number of *N. nitrosa*-type AOB were in a range of $10^5 \sim 10^6$ cells/L. This AOB concentration is very low when compared

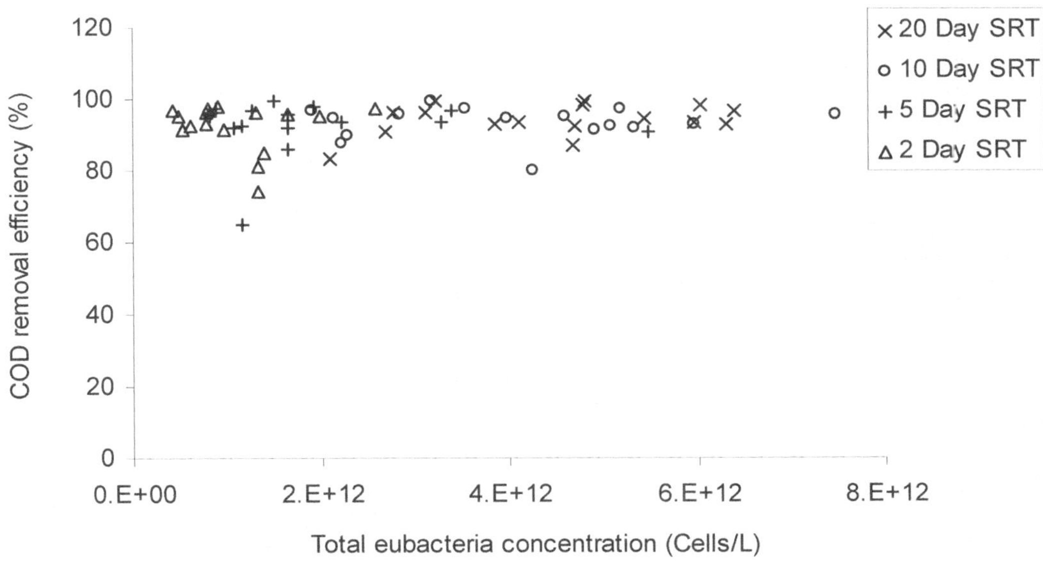

Figure 4-53. Correlation Between Total Eubacteria Concentration and COD Removal Efficiency.

Table 4-62. Quantification of *N. nitrosa*-type AOB (cells/L) in the Industrial Bench-scale System Using a Real-time PCR Assay.

	Phase 1	Phase 2	Phase 3	Phase 4
20-day	4.50 (±2.79) × 10^6	7.13 (±2.49) × 10^6	6.79 (±5.70) × 10^6	6.20 (±6.20) × 10^6
10-day	6.88 (±3.50) × 10^6	7.45 (±3.27) × 10^6	7.27 (±3.56) × 10^6	7.26 (±3.97) × 10^6
5-day	7.33 (±3.24) × 10^6	7.80 (±3.94) × 10^6	17.20 (±4.20) × 10^6	26.60 (±15.80) × 10^6
2-day	0.14 (±0.08) × 10^6	6.81 (±3.73) × 10^6	3.45 (±3.33) × 10^6	0.73 (±0.44) × 10^6

to the range of $10^9 \sim 10^{10}$ cells/L found in earlier full-scale industrial samples. It appeared that *N. nitrosa* was no longer the predominant AOB in the industrial treatment system. Since the bench-scale system continued to effectively nitrify the incoming waste stream (especially in the longer SRT reactors) (Figure 4-54), an AOB population shift must have occurred in the bench-scale system. An assessment of relative population indicated that the number of *N. nitrosa*-type AOB was so low (<0.005% of the total bacteria measured) that this group of organisms was no longer responsible for the level of ammonia oxidation measured in the industrial bench-scale system. Additionally, the correlation between *N. nitrosa* concentration and ammonia-oxidation efficiency was not very strong (Figure 4-55).

Although the *N. nitrosa*-type bacteria were not the major AOB species in the industrial bench-scale system, the population of *N. nitrosa* increased in all four trains as influent ammonia concentrations were increased during phases 2 and 3. Additionally, the *N. nitrosa* population was not affected by the low DO during Phase 3.

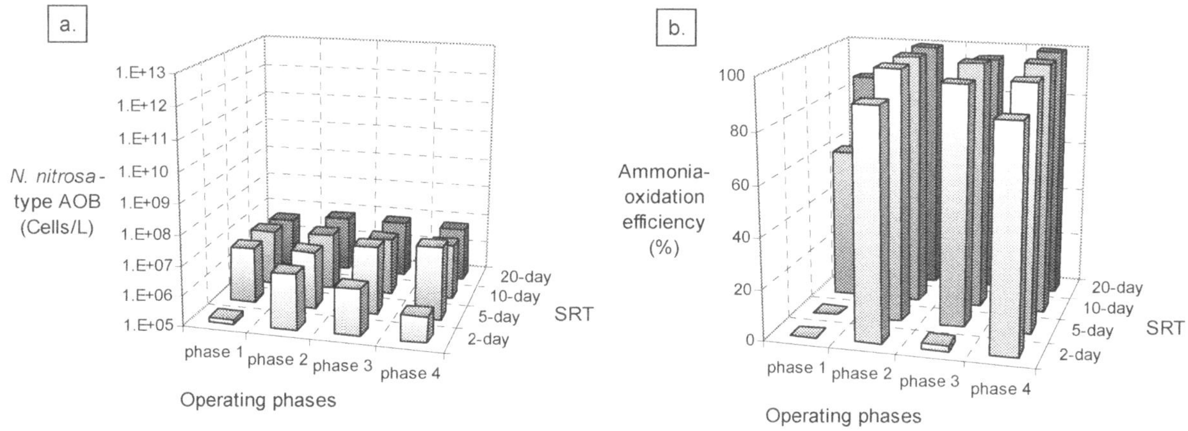

Figure 4-54. *Nitrosomonas nitrosa*-type Ammonia-Oxidizing Bacteria Concentration (cells/L) (A) Compared with the Ammonia-Oxidation Efficiency (%) During Steady State (B) in the Industrial Bench-scale System.

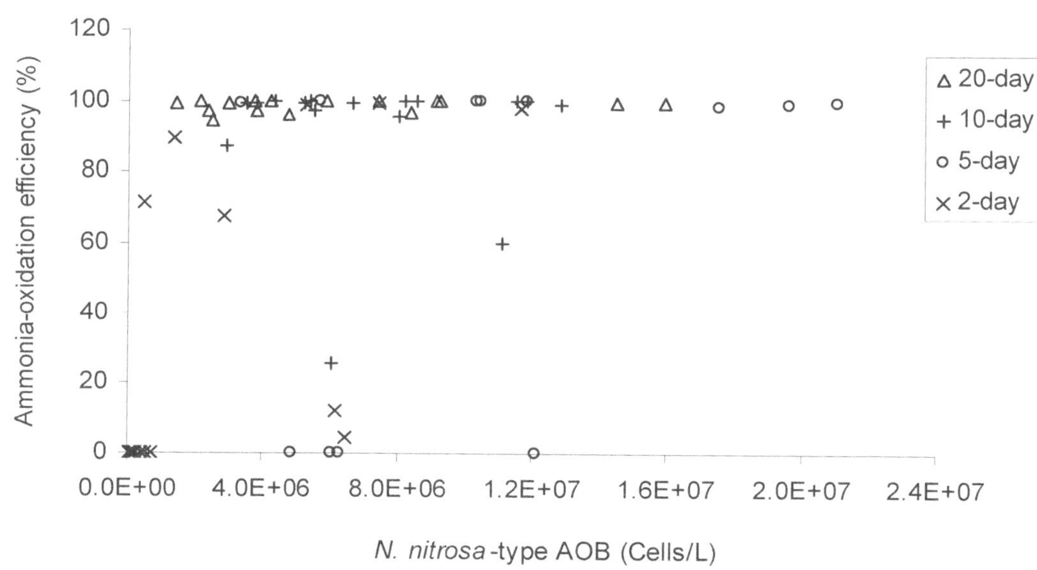

Figure 4-55. Correlation Between *N. nitrosa*-Type AOB Concentration and AOE.

4.2.6 *Nitrospira*-type Nitrite-Oxidizing Bacteria

The *Nitrospira*-type NOB were monitored as the dominant NOB in the industrial bench-scale system. Real-time PCR results are displayed in Table 4-63. The data indicate that the concentration of *Nitrospira*-type NOB in longer SRT trains (10- and 20-day) were higher than those in shorter SRT trains (2- and 5-day) during all four phases. During Phase 1, when nitrite-oxidation did not significantly occur in the 2- and 5-day trains, average *Nitrospira* numbers of 1×10^9 and 1.2×10^9 cells/L were detected in these two trains (Table 4-59). This indicates that

Table 4-63. Quantification of *Nitrospira*-Type Nitrite-Oxidizing Bacteria (cells/L) in the Industrial Bench-scale System Using Real Time-PCR Amplified 16S rDNA.

	Phase 1	Phase 2	Phase 3	Phase 4
20-day	2.59 (±1.68) × 10^{10}	11.10 (±9.06) × 10^{10}	46.70 (±32.40) × 10^{10}	16.50 (±2.02) × 10^{10}
10-day	2.22 (±0.70) × 10^{10}	5.22 (±4.00) × 10^{10}	16.90 (±10.40) × 10^{10}	10.30 (±2.16) × 10^{10}
5-day	0.12 (±0.06) × 10^{10}	2.11 (±1.36) × 10^{10}	2.08 (±1.91) × 10^{10}	3.56 (±1.86) × 10^{10}
2-day	0.10 (±0.07) × 10^{10}	0.18 (±0.03) × 10^{10}	0.17 (±0.17) × 10^{10}	0.38 (±0.09) × 10^{10}

Nitrospira-type NOB can survive low nitrite conditions (during Phase 1) without expressing significant nitrite oxidation activity.

During Phase 2, the concentration of *Nitrospira*-type NOB increased in all four SRT trains, especially in the 5-, 10- and 20-day SRT systems. Nitrite-oxidation performance remained high in the 10- and 20-day SRT trains, and significantly improved in the 2- and 5-day SRT trains (Figure 4-56). This demonstrated that *Nitrospira*-type NOB had been active and proliferated when additional nitrite became available after influent ammonia loading was augmented during the second phase.

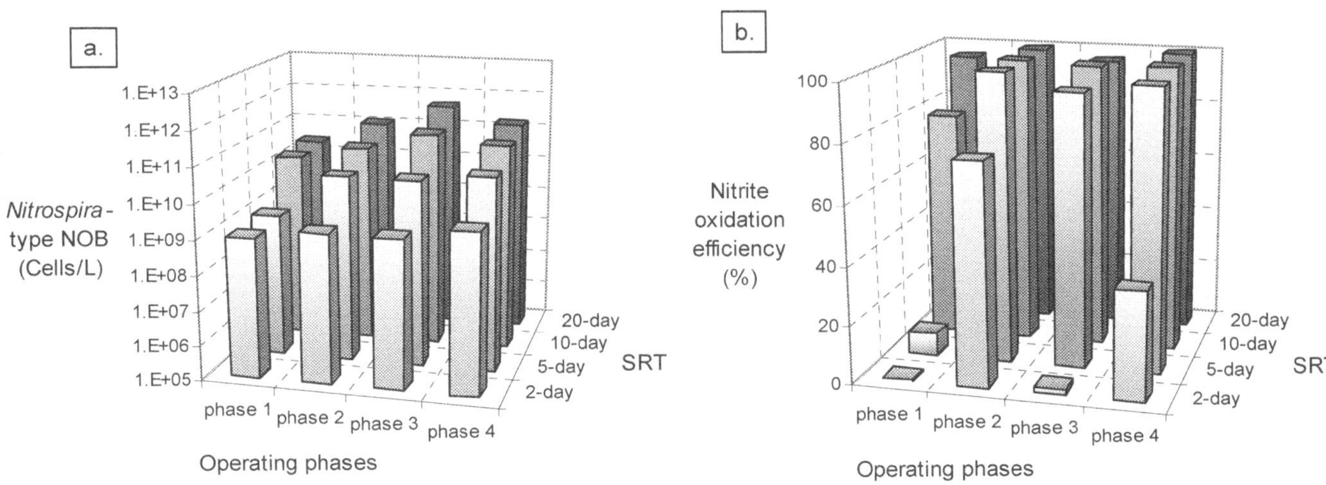

Figure 4-56. ***Nitrospira*-type Nitrite-oxidizing Bacteria Concentration (cells/L) (A) Compared with the Nitrite-oxidation Efficiency (%) During Steady State (B) in the Industrial Bench-scale System.**

During Phase 3 (DO = 0.5 mg/L), *Nitrospira* concentrations in the 2- and 5-day trains did not significantly change while concentrations in the 10- and 20-day trains increased. Low DO operation did not significantly decrease the population of *Nitrospira*-type NOB. Figure 4-57 shows the relationship between *Nitrospira*-type NOB concentration and nitrite-oxidation efficiency. Nitrite-oxidation performance was fairly stable when the *Nitrospira*-type NOB concentration was higher than $1{\times}10^{10}$ cells/L. Instability increased in nitrite oxidation performance when the NOB concentration dropped below this apparent threshold level.

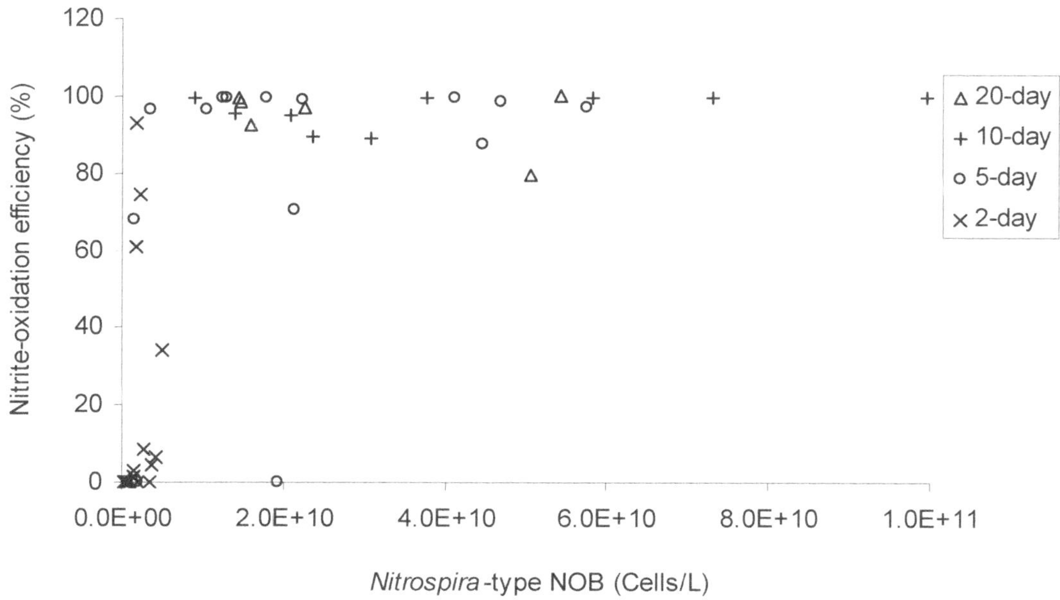

Figure 4-57. Correlation Between *Nitrospira*-type NOB Concentration and Nitrite-oxidation Efficiency.

CHAPTER 5.0

SUMMARY AND CONCLUSIONS

5.1 Municipal Treatment System

5.1.1 Evaluation of Carbon Treatment Performance and Total Biomass

5.1.1.1 COD Removal Efficiency

- Efficient COD removal occurred throughout all phases of the study, including operation at an SRT of 2 days, a temperature of 10°C, and a low DO level.
- Average COD removal efficiencies increased with increasing SRT.
- DO concentration and temperature did not significantly impact COD removal efficiency, which has been attributed to the low $K_{S,O2}$ value associated with heterotrophic bacteria and the high biomass concentrations maintained in activated sludge treatment systems.

5.1.1.2 MLVSS and Total Eubacterial Concentrations

- SRT was the controlling factor in determining the MLVSS concentration, while fluctuations from steady-state could be attributed to changes in influent COD rather than changes in temperature or DO level.
- SRT was found to be the prime determinant of total eubacterial concentration. At temperatures of 20°C and 10°C, no clear trend was observed in total eubacterial cell density in response to changes in DO. However, at a temperature of 15°C, the cell density did increase as the DO level decreased.
- The effect of temperature on the concentration of total eubacteria was ambiguous.
- Trends in total eubacterial cell density were generally similar to those observed in MLVSS for varying operational conditions.

5.1.2 Evaluation of Nitrification Performance

5.1.2.1 Ammonia Oxidation Rate and *N. oligotropha*-type AOB Concentration

- Consistent with the literature, the findings of this study show SRT to be the single most important factor in determining whether or not nitrification will occur in a treatment system. Efficient nitrification occurred at low temperatures and low DO levels for the longer (20- and 10-day) SRT reactors; however, the shorter (5- and 2-day) SRT reactors were dramatically affected by changes in these parameters.

- SRT, DO, and temperature significantly affected ammonia oxidation rates. Based on multiple linear regression analysis, SRT was found to have the largest impact on ammonia oxidation rate, nearly three times temperature and four times DO concentration.
- The concentration of *N. oligotropha*-type AOB and ammonia oxidation rate did not closely follow the same trends with respect to changes in SRT, temperature, and DO, nor were they highly correlated.
- The contribution of *N. oligotropha*-type AOB to the total eubacteria population determined by real-time PCR (≤0.7%) was much lower than expected based on theoretical calculations (5% to 14%). These findings suggest that *N. oligotropha* was not the dominant AOB species in the bench-scale reactors.
- Although, the concentration of *N. oligotropha*-type AOB in the mixed liquor was exceedingly low, the cell density decreased with decreasing temperature and SRT. Statistical analysis indicated little difference between the relative effects of these two parameters. No clear correlation between AOB cell concentration and DO level could be ascertained.

5.1.2.2 Nitrite Oxidation Rate and *Nitrospira*-type NOB Concentration

- Trends in nitrite oxidation rate were very similar to those observed for ammonia oxidation rate.
- SRT, DO, and available nitrite concentration significantly influenced the nitrite oxidation rate. However, the relative impact of available nitrite concentration was more than five times greater than SRT and more than 15 times greater then DO.
- Using real-time PCR analysis, the percentage of *Nitrospira*, relative to total eubacteria, ranged from 0.2% to 5.6% when nitrification occurred. These cell levels were in agreement with the predicted values of 0.4% to 4.0% (based on typical kinetic parameters) suggesting good quantitative agreement between theoretical and measured *Nitrospira* concentrations.
- The SRT and concentration of available nitrite were positively correlated with *Nitrospira* cell densities.
- A strong logarithmic relationship between *Nitrospira* cell density and nitrite oxidation rate was established for all experimental conditions, with a Pearson correlation coefficient of 0.78 between the two measures. When each temperature and DO level were considered separately, the strength of the correlation improved, with coefficients ranging from 0.86 to 0.93.
- A similar relationship was established between the concentration of *Nitrospira* and ammonia oxidation rate, with a correlation coefficient of 0.79 when all experimental conditions were considered and coefficients between 0.84 and 0.95 when each temperature and DO level were considered individually.

- In general, the ammonia oxidation rate controlled both the rate of nitrite oxidation and the *Nitrospira* concentration by regulating the amount of available nitrite. Thus, while SRT, temperature, and DO may directly impact nitrite oxidation and *Nitrospira* concentrations, the indirect effects of these parameters (i.e., their impact on ammonia oxidation) significantly influenced nitrite treatment performance.

5.2 Industrial Treatment System

5.2.1 Evaluation of Carbon Treatment Performance and Total Biomass

5.2.1.1 COD Removal Efficiency

- All four SRT trains of the industrial bench-scale treatment system removed ≥85% of influent COD that ranged in concentration from 486 to 1087 mg/L.
- The average COD removal efficiency in the 20-day SRT train was significantly higher than that of the 2-day SRT treatment train (p-value <0.001).
- Lower DO levels negatively impacted COD removal performance. The average COD removal efficiency during low DO operation was significantly lower than that during the high DO operation (p-value <0.001) at all four SRTs.

5.2.1.2 MLVSS and Total Eubacterial Concentrations

- Total eubacterial concentrations (cells/L) increased when MLVSS levels increased in the longer SRT trains. The linear relationship between the two parameters indicates that higher total eubacteria levels (cells/L) were found in samples with higher concentrations of MLVSS.
- When the total bacteria number was normalized to mass (mg) of MLVSS, the values were essentially constant at approximately 1.31×10^9 cells/mg MLVSS in all four SRT trains during the entire operating period. This indicates that the SRT, DO level, influent ammonia concentration, and COD concentration do not significantly impact the cell density to VSS ratio.

5.2.2 Evaluation of Nitrification Performance

- In general, average TKN removal efficiencies increased with increasing SRT. Little nitrification occurred during high DO-low (unamended) ammonia operation in the 2- and 5-day SRT systems due to heterotrophic uptake of the influent nitrogen.

5.2.2.1 *N. nitrosa*-type AOB Concentration

- The correlation between *N. nitrosa*-type AOB concentration and ammonia-oxidation efficiency was not strong.
- The number of *N. nitrosa*-type AOB measured in the nitrifying bench-scale treatment system ($10^5 \sim 10^6$ cells/L) was significantly lower than that measured in the full-scale WWTP ($10^9 \sim 10^{10}$ cells/L) indicating that an AOB population shift occurred in the laboratory-based system.
- The contribution of *N. nitrosa*-type AOB to the total eubacteria population determined by real-time PCR (< 0.005%) indicated that this group of organisms was no longer

responsible for the level of ammonia oxidation measured in the industrial bench-scale system.

- Although the *N. nitrosa*-type bacteria were not the major AOB species in the industrial bench-scale system, the average concentration of this population increased at all four SRTs when the influent ammonia concentration increased.

5.2.2.2 *Nitrospira*-type NOB Concentration

- Average concentrations of *Nitrospira*-type NOB in longer SRT trains (10- and 20-day) were higher than those in shorter SRT trains (2- and 5-day) during all treatment phases. Additionally, growth of *Nitrospira*-type NOB occurred when additional nitrite was added, although the increase was less pronounced at the lowest (2-day) SRT.
- Low DO operation did not significantly decrease the population of *Nitrospira*-type NOB.
- The *Nitrospira*-type NOB were likely the dominant NOB in the industrial bench-scale system. Using real-time PCR analysis, the percentage of *Nitrospira*, relative to total eubacteria, generally ranged from 0.1 to 4% when nitrification occurred.
- Nitrite-oxidation performance was very good and fairly stable when the *Nitrospira*-type NOB concentration was higher than 1×10^{10} cells/L. Instability increased in nitrite oxidation performance when the NOB concentration dropped below this apparent threshold level.

5.3 Overall Implication

The real-time PCR assay for *Nitrospira* proved to be a useful tool in quantifying the nitrifying population in activated sludge. The concentration of *Nitrospira* measured with this assay was consistent with the NOB cell densities reported in other studies and the percentage of the total population comprised of *Nitrospira* agreed well with predicted values. In addition, *Nitrospira* concentration was found to be highly correlated with both nitrite and ammonia oxidation rates. This measure's close relationship to the ammonia oxidation rate suggests that it may not be necessary to monitor both AOB and NOB, but rather only NOB. This would eliminate the need to characterize the AOB population, which would be beneficial since the high number of AOB species and apparent frequency in population shifts make this a complicated task. This would also reduce the labor and cost associated with monitoring nitrifying organisms in a treatment system.

REFERENCES

Åakra, A.; Utåker, J.B.; Nes, I.F. *FEMS Microbiol. Lett.* **2001**, *205*, 237-242.

Amann, R.I.; Ludwig, W.; Schleifer, K-H. *Microbiol. Rev.* **1995**, *59*, 143-169.

American Public Health Association, *Standard methods for the Examination of Water and* Wastewater, 2nd ed., American Public Health Association, New York, NY, 1998.

Antoniou, P.; Hamilton, J.; Koopman, B.; Jain R.; Holloway B.; Lyberatos, G.; Svoronos S.A. *Water Res.* **1990**, *24*, 97-101.

Applegate, B.; Matrubutham U.; Sanseverino, J.; Sayler, G. *Molecular Microbial Ecology Manual,* **1995**, 6.1.8:1-14.

Atlas, R.M.; Sayler, G.; Burlage, R.; Bej, A. *BioTechniques* **1992**, *12*, 706-716.

Balmelle, B.; Nguyen, K.; Capdeville, B.; Cornier, J.; Deguin, A. *Wat. Sci. Tech.* **1992**, *26*, 1017-1025.

Bassler, H.A.; Flood, S.J.; Livak, K.J.; Marmaro, J.; Knorr, R.; Batt, C.A. *Appl. Environ. Microbiol.* **1995**, *61*, 3724-3728.

Becker, S.; Fahrbach, M.; Böger, P.; Ernst A. *Appl. Environ. Microbiol.* **2002**, *68*, 4486-4494.

Belser, L.W. *Ann. Rev, Micobiol.* **1979**, *33*, 309-333.

Benefield, L.D.; Randall, C.W.; King, P.H. *Temperature Considerations in the Design and Control of Completely-Mixed Activated Sludge Plants*, paper presented at the 2nd Annual National Conference on Environmental Engineering Research, Development and Design, ASCE, University of Florida, Gainesville, FL, 1975.

Benefield, L.D.; Randall, C.W. *Biological Process Design for Wastewater Treatment*, Prentice-Hall: Englewood Cliffs, 1980.

Bernet, N.; Dancong, P.; Delgenes, J-P.; Moletta, R. *Jour. Environ. Eng.* **2001**, *127*, 266-271.

Biesterfeld, S.; Figueroa, L.; Hernandez, M.; Russell, P. *Water Environ. Res.* **2001**, *73*, 329-338.

Biesterfeld, S.; Figueroa, L. *Water Environ. Res.* **2002**, *74*, 470-479.

Bitton, G. *Wastewater Microbiology*, 2nd ed., Wiley-Liss: New York, 1999.

Bjerrum, L.; Kjaer, T.; Ramsing, N.B. *Jour. Microbiol. Methods* **2002**, *51*, 227-239.

Bowers, H.A.,; Tengs, T.; Glasgow, H.B. Jr.; Burkholder, J.M.; Rublee, P.A.; Oldach, D.W. *Appl. Environ. Microbiol.* **2000**, *66*, 4641-4648.

Broude, N.E.. *Trends Biotechnol.* **2001**, *6*, 249-256.

Burrell, P.C.; Kelle,r J.; Blackall, L.L. *Appl. Environ. Microbiol.* **1998**, *64*, 1878-1883.

Cayouette, M.; Sucharczuk, A.; Moores, J.; Tyagi,S.; Kramer F.R. *Strategies Newsletter* **1999**, *12*, 85-88.

Chandler, D.P. *J. Ind. Microbiol. Biotechnol.* **1998**, *21*, 128-140.

Characklis, W.G.; Gujer, W. *Temperature dependency of microbial reactions*. In Kinetics of Wastewater Treatment, SH Jenkins ed., Pergamon Press:Elmsford, NY, pp. 111-130, 1979.

Charley, R.C., Hooper, D.G.; McLee, A.G. *Water Res.* **1980**, *14*, 1387-1396.

Cheremisinoff, N.P. *Biotechnology for Waste and Wastewater Treatment*, Noyles: New Jersey, 1996.

Chuang, S.H.; Ouyang, C.F.; Yuang, H.C.; You, S.J. *Wat. Sci. Tech.* **1997**, *36*, 19-27.

Condon, D.; Liveris, D.; Squires, C.; Schwartz, I.; Squires, C.L. *J. Bacteriol.* **1995**, *177*, 4152.

Cook, K.C.; Layton, A.C.; Dionisi, H.M.; Sayler, G.S. *Abstracts of Papers*, 100th General Meeting of the American Society of Microbiology, N-35, 2000.

Corless, C.E.; Guiver, M.; Borrow, R.; Edwards-Jones, V.; Kaczmarski E.B.; Fox A.J. *J. Clin. Microbiol.* **2000**, *38*, 1747-1752.

Daims, H.; Purkhold, U.; Bjerrum, L.; Arnold, E.; Wilderer, P.A.; Wagner M. *Wat. Sci. Tech.* **2001a**, *43*, 9-18.

Daims, H.; Nielsen, J.L.; Nielsen, P.H.; Schleifer, K.H.; Wagner M. *Appl. Environ. Microbiol.* **2002b**, *67*: 5273-5284.

Dancong, P.; Bernet, N.; Delgenes, J-P.; Moletta, R. *Water Environ. Res.* **2000**, *72*, 195-200.

Davis, M.L.; Cornwell, D.A. *Introduction to Environmental Engineering*, 3rd ed., WCB/McGraw-Hill: Boston, 1998.

Dionisi, H.M.; Layton, A.C.; Harms, G.; Gregory, I.R.; Robinson, K.G.; Sayler, G.S. *Appl. Environ. Microbiol.* **2002a**, *68*, 245-253.

Dionisi, H.M.; Layton, A.C.; Robinson, K.G.; Brown, J.R.; Gregory, I.R.; Parker, J.J.; Sayler, G.S. *Water Environ. Res.* **2002b**, *74*, 462-469.

Dold, P.L.; Wentzel, M.C.; Billing, A.E.; Ekama, G.A.; Marais, G.R. Activated Sludge System Simulation Programs Nitrification and Nitrification/Denitrification Systems Version 1.0, 1991.

Eckenfelder, W.W. Jr. *Industrial Water Pollution Control*. 2nd ed. McGraw-Hill: New York, 1989.

Ehrich S.; Behrens, D.; Lebedeva, E.; Ludwig, W.; Bock, E. *Arch. Microbiol*. **1995**, *164*, 16.

Environmental Protection Agency, Process *Design Manual for Nitrogen Control*. Office of Technology Transfer, Cincinnati, OH, 1975.

Erlich, H.A.; Gelfand, D.; Sninsky, J.J. *Science* **1991**, 252, 1643-1651.

Ferris, M.J.; Muyzer, G.; Ward, D.M. *Appl. Environ. Microbiol*. **1996**, *62*, 340-346.

Fillos, J.; Diyamandoglu, V.; Carrio, L.; Robinson, L. *Water Environ. Res.* **1996**, *68*, 132-142.

Fleming, J.T.; Sanseverino, J.; Sayler, G. *Environ. Sci. Technol.* **1992**, *27*, 1068-1074.

Furumai, H.; Rittmann, B.E. *Wat. Sci. Technol.* **1992**, *26*, 493-502.

Garrett, M.T.; Sawyer, C.N. Proceedings, Seventh Industrial Waste Conference. Lafayette, IN: Purdue University, 1951.

Grady, C.L., Jr.; Daigger, G.T.; Lim, H.C. *Biological Wastewater Treatment*, 2nd ed., Marcel Dekker: New York, 1999.

Grady, C.L.; Filip, C.M. *Water Air Soil Pollut*. **2000**, *123*, 117-132.

Giulietti, A.; Overbergh, L.; Valckx, D.; Decallonne, B.; Boillon, R.; Mathieu C. *Methods* **2001**, *25*, 386-401

Gray, N.F. *Biology of Wastewater Treatment*, Oxford University: New York, 1989.

Grüntzig, V.; Nold, S.C.; Zhou, J.; Tiedje J.M. *Appl. Environ. Microbiol*. **2001**, *67*, 670-768.

Hall, E.R.; Murphy, K.L. *Jour. Water Poll. Control Fed.* **1975**, *57*, 413-418.

Hall, E.R; Murphy, K.L. *Water Res.* **1980**, *14*, 297-304.

Hall, S.J.; Hugenholtz, P.; Slyambaiapitiya, N.; Keller, J.; Blackall, L.L. *Water Sci. Technol.* **2002**, *46*, 267-272.

Hanaki, K.; Wantawin, C.; Ohgaki, S. *Water Res*. **1990**, *24*, 297-302.

Harms, G.; Layton, A.C.; Dionisi, H.M.; Gregory, I.R.; Garrett, V.M.; Hawkins, S.A.; Robinson, K.G.; Sayler G.S. *Environ. Sci. Technol*. **2003**, *37*, 343-351.

Hawkins, S. M.S. Thesis, University of Tennessee at Knoxville, 2000.

Heid,C.A.; Stevens, J.; Livak, K.J.; Williams, P.M. *Genome Res*. **1996**, *6*, 986-994.

Henze, M.; Grady, C.L., Jr.; Gujer, W.; Marais, G.R.; Matsuo, T. *Water Res*. **1987**, *21*, 505-515.

Henze, M.; Harremoes, P.; Jansen, J.C.; Arvin, E. *Wastewater Treatment: Biological and Chemical Processes*, Springer-Verlag: Berlin Heidelberg, 1995.

Hermansson, A.; Lindgren, P. *Appl. Environ. Microbiol*. **2001**, *67*, 972-976

Higuchi, R.; Fockler, C.; Dollinger, G.; Watson, R. *Bio/Technology*. **1993,** *11,* 1026-1030.

Hillis, D.; Allard, M.; Miyamoto, M. *Methods in Enzymology* **1993**, *224*, 456-487.

Holland, P.M.; Abramson, R.D.; Watson, R.; Gefand, D.H. *Proc. Natl. Acad. Sci. USA*. **1991**, *88*, 7276-7280.

Hovanec, T.A.; Delong, E.F. *Appl. Environ. Microbiol*. **1996**, *62*, 2888.

Hovanec, T.A.; Taylor, L.T.; Blakis, A.; Delong, E.F. *Appl. Environ. Microbiol*. **1998**, *64*, 258-264.

Jenkins, D.; Richard, M.G.; Daigger, G.T.; *Manual on the Causes of Control of Activated Sludge Bulking and Foaming*. 2nd ed. Lewis Publishers, Chelsea, MI, 1993.

Jialong, W.; Hanchang, S.; Yi, Q. *Process Biochem*. **2000**. *36*, 297-303.

Johnsen, K.; Enger, O.; Jacobsen, C.S.; Thirup, L.; Torsvik V. *Appl. Environ. Microbiol.* **1999**, *65*, 1786-1789.

Juretschko, S.;Timmermann, G; Schmid, M.; Schleifer, K.; Pommerening-Roser, A.; Koops, H.; Wagner, M. *Appl. Environ. Microbiol*. **1998**, *64*, 3042-3051.

Kappeler, J.; Gujer, J. *Wat. Sci. Tech*. **1994**, *30*, 181-189.

Keller, J.; Yuan, Z.; Blackall, L.L. *Environ. Sci. Biotechnol*. **2002**, *1*, 83-97.

Kim, I.S.; Ivanov, V.N. *World J. Microbiol. Biotechnol*. **2000**, *16*, 425-430.

Klappenbach, J.A.; Dunbar, J.M.; Schmidt, TM. *Appl. Environ. Microbiol*. **2000**, *66*, 1328.

Klappenbach, J.A.; Saxman, P.R.; Cole, J.R., Schmidt, T.M. *Nucleic Acids Res*. **2001**, *29*, 181-184.

Knowles, G.; Downing, A.L.; Barrett, M.J. *Jour. General Microbiol.* **1965**, *38*, 263-276.

Kostrikis, L.G.; Tyagi,S.; Mhlanga, M.M.; Ho, D.D.; Kramer, F.R. *Science* **1998**, *279*, 1228-1229.

Kowalchuk, G.; Stephen, J.; De Boer, W.; Prosser, J.; Embley, T.; Woldendorp, J. *Appl. Environ. Microbiol.* **1997**, *63*, 1489-1497.

Kowalchuk, G.A.; Naoumenko, Z.S.; Derikx, P.L.; Felske, A.; Stephen, J.R.; Arkhipchenko, I.A.; *Appl. Environ. Microbiol*. **1999**, *65*, 396-403.

Kowalchuk, G.A.; Stephen, J.R. *Annu. Rev. Microbiol*. **2001**, *55*, 485-529

Laanbroek, H.J.; Bodelier, P.E.; Gerards, S. *Arch. of Microbiol*. **1994**, *161*, 156-162.

Lane, D.J. In *Nucleic Acid Techniques in Bacterial Systematics*, Stackebrandt, E.; Goodfellow, M., Ed., John Wiley & Sons, Ltd., 1991.

Lawrence, A.W.; Brown, C.G. *Jour. Water Poll. Control Fed.* **1976**, *48*, 1179-1802.

Lawrence, A.; McCarty, P. *Journal of the Proceedings of the American Society of Civil Engineers* **1970**, *96*, 757-778.

Lau, A.O.; Strom, P.F.; Jenkins, D. *Jour. Water Poll. Control Fed.* **1984**, *56*, 41-51.

Lee, H.S.; Park, S.J.; Yoon, T.I. *Process Biochem*. **2002**, *38,* 81-88

Lee, L.G.; Connell, C.R.; Bloch, W. *Nucleic Acids Res*. **1993**, *21*, 3761-3766.

Lee, S-Y.; Bollinger, J.; Bezdicek, D.; Ogram, A. *Appl. Environ. Microbiol.* **1996**, *62*, 3787-3793.

Lie, Y.S.; Petropoulos, C.J. *Curr. Opin. Biotech*. **1998**, *9*, 43-48.

Lishman, L.A.; Legge, R.L.; Farquhar, G.J. *Water Res.* **2000**, *34*, 2263-2276.

Livak, K.J.; Flood, S.A.; Marmaro, J.; Giusti, W.; Deetz, K. *PCR Methods Applications* **1995a**, *4*, 357-362.

Livak, K.J.; Marmaro, J.; Todd, J.A. *Nature Genet*. **1995b**, *9*, 341-342.

Madigan, M.T.; Martinko, J.M.; Parker, J. *Brock Biology of Microorganisms.* 8th ed.; Prentice-Hall: New York, 1997.

Marisili-Libelli, S.; Giovanni, F. *Water Res*. **1997**, *31*, 179-185.

McClintock, S.A.; Randall, C.W.; Pattarkine, V.M. *Water Environ. Res.* **1993**, *65*, 110-118.

McTavish, H.; Fuchs, J.A.; Hooper, A.B. *J. Bacteriol.* **1993**, *175*, 2436-2444.

Mendum, T.A.; Sockett, R.E.; Hirsch, P.R. *Appl. Environ. Microbiol.* **1999**, *65*, 4155-4162.

Metcalf and Eddy, *Wastewater Engineering: Treatment, Disposal, Reuse*, 3rd ed. McGraw-Hill: New York, 1991.

Mobarry, B.K.; Wagner, M.; Urbain, V.; Rittmann, B.E.; Stahl, D. *Appl. Environ. Microbiol.* **1996**, *62*, 2156-2162.

Mobarry, B.K.; Wagner, M.; Urbain, V.; Rittmann, B.E.; Stahl, D. *Appl. Environ. Microbiol.* **1997**, *63*, 815.

Muck, R.E.; Grady, C.L., Jr. *Temperature effects on microbial growth in CSTRs*. Journal of the Environmental Engineering Division, ASCE 100:1147-1163, 1974.

Munch, E.V.; Barr, K.; Watts, S.; Keller, J. *Wat. Sci. Tech.* **2000**, *41*, 5-12.

Navarro, E.; Fernandez, M.P.; Grimont, F.; Clays-Josserand, A.; Bardini, R. *Int. J. Syst. Bacteriol.* **1992**, *42*, 554-560.

Ng ,W.J.; Yap, M.S.; Sivadas, M. *Biological Wastes* **1989**, *29*, 299-311.

Norton, J.M.; Alzerreca, J.J.; Suwa, Y.; Klots, M.G. *Arch. Microbiol.* **2002**, *177*, 139-149.

Norusis, M.J. *SPSS 10.0 Guide to Data Analysis.* Prentice-Hall: Upper Saddle River, NJ, 2000.

Oerther, D.B.; de los Reyes, F.L.; Rankin, L. *Wat. Sci. Technol.* **1999**, *39*, 1-20.

Okabe, S.; Satoh, H.; Watanabe, Y. *Appl. Environ. Microbiol.* **1999**, *65*, 3182-3191.

Olsen, G.J.; Lane, D.J.; Giovannoni, S.J.; Pace, N.R.; Stahl, D.A. *Annu. Rev. Microbiol.* **1986,** *40*, 337-365.

Palm, J.C.; Jenkins, D.; Parker, D.S. *Jour. Water Poll. Control Fed.* **1980**, *52*, 2484-2506.

Phillips, C.J.; Paul, E.A.; Prosser, J.I. *FEMS Microbiol. Ecol.* **2000**, *32*, 167-175.

Piatek, A.S.; Tyagi, S.; Pol, A.C.; Telenti, A.; Miller, L.P.; Kramer, F.R.; Alland, D. *Nature Biotechnology* **1998**, *16*, 359-363.

Poduska, R.A.; Andrews, J.F. *Jour. Water Poll. Control Fed.* **1975**, *47*, 2599-2619.

Princic, A.; Mahne, I.; Megušar, F.; Paul, E.A.; Tiedje, J.M. *Appl. Environ. Microbiol.* **1998**, *64*, 3584-3590.

Prosser, J.I. *Adv. Microb. Physiol.* **1989**, *30*, 897-925.

Purkhold, U.; Pommerening-Röser, A.; Juretschko, S.; Schmid, M.C.; Koops, H.P.; Wagner, M. *Appl. Environ. Microbiol.* **2000**, *65*, 5368-5382.

Randall, C.W.; Buth, D. *Jour. Water Poll. Control Fed.* **1984**, *56*, 1039-1044.

Reynolds, T.D.; Richards, P.A. *Unit Operations and Processes in Environmental Engineering*, 2nd ed., PWS Publishing: Boston, 1996.

Rittmann, B.E.; Laspidou, C.S.; Flax, J.; Stahl, D.A.; Urbain, V.; Harduin, H.; van der Waarde, J.J., Geurkink, B.; Henssen, M.C.; Brouwer, H.; Klapwijk, A.; Wetterauw, M. *Wat. Sci. Technol.* **1999**, *39*, 51-59.

Rittmann, B.E.; McCarty, P.L. *Environmental Biotechnology: Principles and Applications*, McGraw-Hill: New York, 2001.

Rittmann, B.E. *Water Environ. Res.* **2002**, *74*, 421-427.

Rotthauwe, J.; Witzel, K.; Liesack, W. *Appl. Environ. Microbiol.* **1997**, *63*, 4704-4712.

Saiki, R.K.; Scharf , S.; Faloona, F.; Mullis, K.B.; Horn, G.T.; Erlich, H.A.; Arnheim, N. *Science* **1985**, *230*, 1350-1354.

Saiki, R.K.; Gelfand, D.H.; Stoffel, S.; Scharf , S.J.; Higuchi, R.; Horn, G.T.; Mullis, K.B; Erlich, H.A. *Science* **1988**, *239*, 487-491.

Sambrook, J.; Fritsch, E.F.; Maniatis, T. *Molecular cloning: a laboratory manual*, 2nd ed. Cold Harbor Laboratory, Cold Spring Harbor, NY, 1989.

Sayigh, B.A.; Malina, J.F. *Jour. Water Poll. Control Fed.* **1978**, *50*, 678-687.

Schramm, A.; Larsen, L.J.; Revsbech, N.P.; Ramsing, N.B.; Amann, R.; Schleifer, K-H. *Appl. Environ. Microbiol.* **1996**, *62*, 4641.

Schramm, A.; de Beer, D.; Wagner, M.; Amann, R. *Appl. Environ. Microbiol.* **1998**, *64*, 3480-3485.

Schramm, A.; de Beer, D.; Van Den, Heubel J.C.; Ottengraf, S.; Amann R. *Appl. Environ. Microbiol.* **1999**, *65*, 3690-3696.

Shammas, N. *Jour. Water Poll. Control Fed.* **1986**, *58*, 52-59.

Sinclair, C.G.; Ryder, D.N. *Biotechnol. Bioengineer.* **1975**, *17*, 375-398.

Stenstrom, M.; Song, S. *Jour. Water Poll. Control Fed.* **1991**, *63*, 208-219.

Stephen, J.R.; Chang, Y-J.; Macnaughton, S.J.; Kowalchuk, G.A.; Leung, K.T., Flemming, C.A.; White, D.C. *Appl. Environ. Microbiol.* **1999**, *65*, 95-101.

Strover, E.L.; Kincannon, D.F. *Jour. Water Poll. Control Fed.* **1976**, *48*, 645-651.

Sucuru, G.; Cetin, F. *Environ. Technol.* **1990**, *11*, 205-212.

Suzuki, M.T.; Taylor, L.T.; Delong, E.F. *Appl. Environ. Microbiol.* **2000**, *66*, 4605-4614.

Suzuki, M.T.; Taylor, L.T.; Delong, E.F. *Appl. Environ. Microbiol.* **2001**, *67*, 972-976.

Tchobanoglous, G.; Burton F.L. *Wastewater Engineering: Treatment, Disposal, and Reuse*, 3rd ed., Irwin, McGraw-Hill, 1991.

Tennessee Department of Public Health. *Rules of the Tennessee Department of Public Health Bureau of Environmental Health Services Division of Water Quality Control.* Nashville, 1999.

Thompson, J.D.; Higgens, D.G.; Gibson, T.J. *Nucleic Acids Res.* **1994**, *22*, 4673-4680.

Tortora, G.J.; Funke, B.R.; Case, C.L. Microbiology: An Introduction, 3rd ed., Benjamine/Cummings: Redwood City, 1993.

Tyagi, S.; Bratu, D.P.; Kramer, F.R. *Nature Biotechnology* **1998**, *16*, 49-53.

Tyagi, S.; Kramer, F.R. *Nature Biotechnology* **1996**, *14*, 303-308.

Tyagi, S.; Marra, S.E.; Vet, J.M.; Kramer, F.R. http://molecular-beacons.org 1999.

Urbain,V.; Mobarry, B.; de Silva, V.; Stahl, D.A.; Rittmann, B.E.; Manem, J. *Wat. Sci. Technol.* **1998**, *37*, 223-229.

Vet, J.M.; Majithia,A.R.; Marras, S.E.; Tyagi, S.; Dube, S.; Poiesz, B.J.; Kramer, F.R. *Proc. Natl. Acad. Sci. USA*, **1999**, *96*, 6394-6399.

Wagner, M.; Amann, R.; Lemmer, H.; Schleifer, K-H. *Appl. Environ. Microbiol.* **1993**, *59*, 1520-1525.

Wagner, M; Rath, G.; Amann, R.; Koops, H-P.; Schleifer, K-H. *System. Appl. Microbiol.* **1995**, *18*, 251-264.

Wagner, M.; Rath, G.; Koops, H-P.; Flood, J.; Amann, R. *Water Sci. Technol.* **1996**. *34*, 237-244.

Wagner, M.; Noguera, D.R.; Juretschko, S.; Rath, G.; Koops, H.; Schleifer, K. *Wat. Sci. Technol.* **1998**, *37*, 441-449.

Wagner, M.; Loy, A.; Nogueira, R.; Purkhold, U.; Lee, N.; Daims, H. *Anton. Leeuwenhoek Int. J. Gen. M.* **2002**, *81*, 665-680.

Wanner, J. *J. Inst. Water Environ. Management* **1998**, *12*, 368-374.

Wild, H.E. Jr.; Sawyer, C.N.; McMahon, T.C. *Jour. Water Poll. Control Fed.* **1971**, *43*, 1845-1854.

Wilderer, P.A.; Bungartz, H.J.; Lemmer, H.; Wagner, M.; Keller J.; Wuertz, S. *Water Res.* **2002**, *36*, 370-393.

Wong-Chong, G.M.; Loehr, R.C. *Water Res.* **1975**, *9*, 1099-1106.

Ydstebo, L.; Bilstad, T.; Barnard, J. *Water Environ. Res.* **2000**, *72*, 444-454.

You, S.J.; Hsu, C.L.; Chuang, S.H.; Ouyang C.F. *Water Res.* **2003**, *37*, 2281-2290.

Yuan, Z.; Blackall, L.L. *Water Res.* **2002**, *36*, 482-490.

Zheng, H.; Hanaki, K.; Matsuo, T. *Wat. Sci. Tech.* **1994**, *30*, 91-100.

WASTEWATER UTILITY

Alabama
Montgomery Water Works & Sanitary Sewer Board

Alaska
Anchorage Water & Wastewater Utility

Arizona
Gila Resources

Glendale, City of, Utilities Department

Mesa, City of

Peoria, City of

Phoenix Water Services Department

Pima County Wastewater Management

Arkansas
Little Rock Wastewater Utility

California
Calaveras County Water District

Central Contra Costa Sanitary District

Contra Costa Water District

Crestline Sanitation District

Delta Diablo Sanitation District

Dublin San Ramon Services District

East Bay Dischargers Authority

East Bay Municipal Utility District

El Dorado Irrigation District

Fairfield-Suisun Sewer District

Fresno Department of Public Utilities

Irvine Ranch Water District

Las Virgenes Municipal Water District

Livermore, City of

Lodi, City of

Los Angeles, City of

Los Angeles County, Sanitation Districts of

Napa Sanitation District

Orange County Sanitation District

Palo Alto, City of

Riverside, City of

Sacramento Regional County Sanitation District

San Diego Metropolitan Wastewater Department, City of

San Francisco, City & County of

San Jose, City of

Santa Barbara, City of

Santa Rosa, City of

South Bayside System Authority

South Coast Water District

South Orange County Wastewater Authority

Union Sanitary District

West Valley Sanitation District

Colorado
Boulder, City of

Colorado Springs Utilities

Littleton/Englewood Water Pollution Control Plant

Metro Wastewater Reclamation District, Denver

Connecticut
The Mattabasset District

New Haven, City of, WPCA

District of Columbia
District of Columbia Water & Sewer Authority

Florida
Broward, County of

Fort Lauderdale, City of

Gainesville Regional Utilities

JEA

Miami-Dade Water & Sewer Authority

Orange County Utilities Department

Orlando, City of

Reedy Creek Improvement District

Seminole County Environmental Services

St. Petersburg, City of

Stuart Public Utilities

Tallahassee, City of

Tampa, City of

Toho Water Authority

West Palm Beach, City of

Georgia
Atlanta Department of Watershed Management

Augusta, City of

Clayton County Water Authority

Cobb County Water System

Columbus Water Works

Fulton County

Gwinnett County Department of Public Utilities

Macon Water Authority

Savannah, City of

Hawaii
Honolulu, City and County of

Illinois
American Bottoms Wastewater Treatment Plant

Greater Peoria Sanitary District

Kankakee River Metropolitan Agency

Metropolitan Water Reclamation District of Greater Chicago

North Shore Sanitary District

Wheaton Sanitary District

Indiana
Fort Wayne, City of

Iowa
Ames, City of

Cedar Rapids Wastewater Facility

Des Moines Metro Wastewater Reclamation Authority

Iowa City, City of

Kansas
Johnson County Unified Wastewater Districts

Unified Government of Wyandotte County/Kansas City, City of

Kentucky
Louisville & Jefferson County Metropolitan Sewer District

Louisiana
Sewerage & Water Board of New Orleans

Maine
Bangor, City of

Portland Water District

Maryland
Anne Arundel County Bureau of Utility Operations

Howard County Department of Public Works

Washington Suburban Sanitary Commission

Massachusetts
Boston Water & Sewer Commission

Upper Blackstone Water Pollution Abatement District

Michigan
Ann Arbor, City of

Detroit, City of

Holland Board of Public Works

Lansing, City of

Owosso Mid-Shiawassee County WWTP

Saginaw, City of

Wayne County Department of Environment

Wyoming, City of

Minnesota
Rochester, City of

Western Lake Superior Sanitary District

Missouri
Independence, City of

Kansas City Missouri Water Services Department

Little Blue Valley Sewer District

Nebraska
Lincoln Wastewater System

Nevada
Henderson, City of

New Jersey
Passaic Valley Sewerage Commissioners

New York
New York City Department of Environmental Protection

Rockland County Solid Waste Management Authority/ Sewer District

North Carolina
Cary, Town of

Charlotte/Mecklenburg Utilities

Durham, City of

Metropolitan Sewerage District of Buncombe County

Orange Water & Sewer Authority

Ohio
Akron, City of

Butler County Department of Environmental Services

Columbus, City of

Metropolitan Sewer District of Greater Cincinnati

Northeast Ohio Regional Sewer District

Summit, County of

Oklahoma
Tulsa, City of

Oregon
Clean Water Services

Eugene/Springfield Water Pollution Control

Water Environment Services

Pennsylvania
Philadelphia, City of

University Area Joint Authority, State College

South Carolina
Charleston Commissioners of Public Works

Mount Pleasant Waterworks & Sewer Commission

Spartanburg Sanitary Sewer District

Tennessee
Cleveland, City of

Knoxville Utilities Board

Murfreesboro Water & Sewer Department

Nashville Metro Water Services

Texas
Austin, City of

Dallas Water Utilities

Denton, City of

El Paso Water Utilities

Fort Worth, City of

Gulf Coast Waste Disposal Authority

Houston, City of
San Antonio Water System
Trinity River Authority

Utah
Salt Lake City Corporation

Virginia
Alexandria Sanitation Authority
Arlington, County of
Fairfax County Virginia
Hampton Roads Sanitation District
Henrico, County of
Hopewell Regional Wastewater Treatment Facility
Loudoun County Sanitation Authority
Lynchburg Regional WWTP
Prince William County Service Authority
Richmond, City of
Rivanna Water & Sewer Authority

Washington
Edmonds, City of
Everett, City of
King County Department of Natural Resources
Seattle Public Utilities
Sunnyside, Port of
Yakima, City of

Wisconsin
Green Bay Metro Sewerage District
Madison Metropolitan Sewerage District
Milwaukee Metropolitan Sewerage District
Racine, City of
Sheboygan Regional Wastewater Treatment
Wausau Water Works

Australia
South Australian Water Corp.
Sydney Water Corp.
Water Corp. of Western Australia

Canada
Greater Vancouver Regional District
Toronto, City of, Ontario
Winnipeg, City of, Manitoba

Mexico
Servicios de Agua y Drenaje de Monterrey, I.P.D.

New Zealand
Watercare Services Limited

Singapore
Singapore Public Utilities Board

United Kingdom
Yorkshire Water Services Limited

STORMWATER UTILITY

California
Los Angeles, City of, Department of Public Works
Monterey, City of
Santa Rosa, City of

Colorado
Boulder, City of

Georgia
Griffin, City of

Iowa
Cedar Rapids Wastewater Facility
Des Moines Metro Wastewater Reclamation Authority

Kansas
Overland Park, City of

Kentucky
Louisville & Jefferson County Metropolitan Sewer District

Maine
Portland Water District

Minnesota
Western Lake Superior Sanitary District

North Carolina
Cary, Town of
Charlotte, City of, Stormwater Services

Pennsylvania
Philadelphia, City of

Tennessee
Chattanooga Storm Water Management

Washington
Seattle Public Utilities

STATE

Arkansas Department of Environmental Quality
Fresno Metropolitan Flood Control District, Calif.
Urban Drainage & Flood Control District, Colo.

CORPORATE

ADS Environmental Services
The ADVENT Group Inc.
Alan Plummer & Associates
Alden Research Laboratory
Alpine Technology Inc.
Aqua-Aerobic Systems, Inc.
Aquateam–Norwegian Water Technology Centre A/S
BaySaver Inc.
BioVir Laboratories, Inc.
Black & Veatch
Boyle Engineering Corporation
BPR CSO
Brown & Caldwell
Burns & McDonnell
CABE Associates Inc.
The Cadmus Group
Camp Dresser & McKee Inc.
Carollo Engineers Inc.
Carpenter Environmental Associates Inc.
CDS Technologies Inc.
Chemtrac Systems Inc.
CH2M HILL
Damon S. Williams Associates, LLC
David L. Sheridan, P.C.
Earth Tech Inc.
Eco-Matrix
Environmental Engineers International
EMA Inc.
The Eshelman Company Inc.
Finkbeiner, Pettis, & Strout Inc.
Freese & Nichols, Inc.
Frontier Geosciences Inc.
ftn Associates Inc.
Fuss & O'Neill, Inc.
Gannett Fleming Inc.
Golder Associates Inc.
Greeley and Hansen LLC
Hazen & Sawyer, P.C.
HDR Engineering Inc.
HNTB Corporation
HydroQual Inc.
Ingersoll-Rand Energy Systems
Insituform Technologies Inc.
Institute for Environmental Technology & Industry, Korea
Jacobson Helgoth Consultants Inc.
Jason Consultants Inc.
Jordan, Jones, & Goulding Inc.
KCI Technologies Inc.
Kelly & Weaver, P.C.
Kennedy/Jenks Consultants
Komline Sanderson Engineering Corporation
Lawler, Matusky & Skelly Engineers, LLP
Limno-Tech Inc.
Lombardo Associates Inc.
Malcolm Pirnie Inc.
McKim & Creed
MEC Analytical Systems Inc.
Metcalf & Eddy Inc.
MWH
New England Organics
Odor & Corrosion Technology Consultants Inc. (OCTC)
ONDEO Degremont Inc.
Oswald Green, LLC
PA Government Services Inc.
Parametrix Inc.
Parsons
Post, Buckley, Schuh & Jernigan
R&D Engineering/Conestoga Rover & Associates
The RETEC Group
RMC, Inc. (Raines, Melton & Carella)
R.M. Towill Corporation
Ross & Associates Ltd.
Royce Technologies
SAIC Maritime Technical Group
Sear-Brown
Short Elliott Hendrickson, Inc.
Stormwater Management, Inc.
Synagro Technologies, Inc.
Tetra Tech Inc.
Trojan Technologies Inc.
URS Corporation
Veolia Water NATC
Wade-Trim Inc.
Weston Solutions Inc.
Woodard & Curran
WRc/D&B, LLC
WWETCO, LLC
Zoeller Pump Company

INDUSTRY

American Electric Power
ChevronTexaco Energy Research & Technology Company
The Coca-Cola Company
Dow Chemical Company
DuPont Company
Eastman Kodak Company
Merck & Company Inc.
ONDEO Services
Procter & Gamble Company
PSEG Services Corp.
Severn Trent Services Inc.
Shell Global Solutions
Thames Water Plc
United Water Services LLC

Note: List as of 1/31/04

Board of Directors

Chair
James F. Stahl
County Sanitation Districts of Los Angeles County

Vice-Chair
Vernon D. Lucy
ONDEO Degremont Inc.

Secretary
William J. Bertera
Water Environment Federation

Treasurer
Karl W. Mueldener
Kansas Department of Health & Environment

Mary E. Buzby, Ph.D.
Merck & Company Inc.

Dennis M. Diemer, P.E.
East Bay Municipal Utility District

Jerry N. Johnson
District of Columbia Water and Sewer Authority

Richard D. Kuchenrither, Ph.D.
Black & Veatch

Alfonso R. Lopez
New York City Department of Environmental Protection

Richard G. Luthy, Ph.D.
Stanford University

John T. Novak, Ph.D.
Virginia Polytechnic Institute & State University

Lynn H. Orphan
Kennedy/Jenks Consultants

J. Michael Read
HDR, Inc.

James M. Tarpy
Nashville Metro Water Services

Murli Tolaney
MWH

Executive Director
Glenn Reinhardt

Research Council

Chair
John Thomas Novak, Ph.D.
Virginia Polytechnic Institute & State University

Vice-Chair
Glen T. Daigger, Ph.D.
CH2M HILL

Robert G. Arnold, Ph.D.
University of Arizona, Tucson

Robin L. Autenrieth, Ph.D.
Texas A&M University

William L. Cairns, Ph.D.
Trojan Technologies, Inc.

James Crook, Ph.D.
Water Reuse Consultant

Geoffrey H. Grubbs
U.S. EPA

Mary A. Lappin, P.E.
Kansas City Water Services Department

Keith J. Linn
Northeast Ohio Regional Sewer District

Drew C. McAvoy, Ph.D.
Procter & Gamble Company

Margaret H. Nellor, P.E.
County Sanitation Districts of Los Angeles County

Spyros Pavlostathis, Ph.D.
Georgia Institute of Technology

Steven M. Rogowski, P.E.
Metro Wastewater Reclamation District of Denver

Peter J. Ruffier
Eugene/Springfield Water Pollution Control

Michael W. Sweeney, Ph.D.
EMA Inc.

George Tchobanoglous, Ph.D.
Tchobanoglous Consulting

Gary Toranzos, Ph.D.
University of Puerto Rico

Stormwater Technical Advisory Committee

Chair
Robert E. Pitt, Ph.D., P.E., D.E.E.
University of Alabama

Vice-Chair
Ben Urbonas, P.E.
Urban Drainage and Flood Control District

Christine Andersen, P.E.
City of Long Beach, California

Gail B. Boyd
URS Corporation

Larry Coffman
Prince George's County

Brian Marengo, P.E.
City of Philadelphia Water Department

A. Charles Rowney, Ph.D.
Camp Dresser & McKee Inc.

James Wheeler, P.E.
U.S. EPA

WERF Product Order Form

As a benefit of joining the Water Environment Research Foundation, subscribers are entitled to receive one complimentary copy of all final reports and other products.Additional copies are available at cost (usually $10).To order your complimentary copy of a report, please write "free" in the unit price column.WERF keeps track of all orders.If the charge differs from what is shown here, we will call to confirm the total before processing.

Name ________________ Title ________________

Organization ________________

Address ________________

City ________________ State ________ Zip Code ________ Country ________

Phone ________________ Fax ________________ Email ________________

Stock #	Product	Quantity	Unit Price	Total
			Postage & Handling	
			VA Residents Add 4.5% Sales Tax	
			Canadian Residents Add 7% GST	
			TOTAL	

Method of Payment: (All orders must be prepaid.)

❑ Check or Money Order Enclosed

❑ Visa ❑ Mastercard ❑ American Express

Account No. ________________ Exp. Date ________

Signature ________________

Shipping & Handling:

Amount of Order	United States	Canada & Mexico	All Others
Up to but not more than:	Add:	Add:	Add:
$20.00	$5.00*	$8.00	50% of amount
30.00	5.50	8.00	40% of amount
40.00	6.00	8.00	
50.00	6.50	14.00	
60.00	7.00	14.00	
80.00	8.00	14.00	
100.00	10.00	21.00	
150.00	12.50	28.00	
200.00	15.00	35.00	
More than $200.00	Add 20% of order	Add 20% of order	

*minimum amount for all orders

Note: Please make checks payable to the Water Environment Research Foundation.

To Order (Subscribers Only):

Log on to www.werf.org and click on the "Product Catalog."

Phone: (703) 684-2470
Fax: (703) 299-0742.

WERF
Attn: Subscriber Services
635 Slaters Lane
Alexandria, VA 22314-1177

To Order (Non-Subscribers):

Non-subscribers may be able to order WERF publications either through WEF (www.wef.org) or IWAP (www.iwapublishing.com). Visit WERF's website at www.werf.org for details.

IWA Publishing's authorised EU representative for General Product Safety Regulations is Diane D'Arras, 15 rue Duret, 75116 Paris, France, e-mail: safety@iwap.co.uk.

Printed and bound by CPI Group (UK) Ltd, Croydon, CR0 4YY
12/05/2026
02108886-0007